Safety Improvements through Lessons Learned from Operational Experience in Nuclear Research Facilities

NATO Security through Science Series

A Series presenting the results of scientific meetings supported under the NATO Programme for Security through Science (STS).

The Series is published by IOS Press, Amsterdam, and Springer Science and Business Media, Dordrecht, in conjunction with the NATO Public Diplomacy Division.

Sub-Series

A. Chemistry and Biology	(Springer Science and Business Media)
B. Physics and Biophysics	(Springer Science and Business Media)
C. Environmental Security	(Springer Science and Business Media)
D. Information and Communication Security	(IOS Press)
E. Human and Societal Dynamics	(IOS Press)

Meetings supported by the NATO STS Programme are in security-related priority areas of Defence Against Terrorism or Countering Other Threats to Security. The types of meeting supported are generally "Advanced Study Institutes" and "Advanced Research Workshops". The NATO STS Series collects together the results of these meetings. The meetings are co-organized by scientists from NATO countries and scientists from NATO's "Partner" or "Mediterranean Dialogue" countries. The observations and recommendations made at the meetings, as well as the contents of the volumes in the Series, reflect those of the participants in the workshop. They should not necessarily be regarded as reflecting NATO views or policy.

Advanced Study Institutes (ASI) are high-level tutorial courses to convey the latest developments in a subject to an advanced-level audience

Advanced Research Workshops (ARW) are expert meetings where an intense but informal exchange of views at the frontiers of a subject aims at identifying directions for future action

Following a transformation of the programme in 2004 the Series has been re-named and re-organised. Recent volumes on topics not related to security, which result from meetings supported under the programme earlier, may be found in the NATO Science Series

www.nato.int/science
www.springeronline.com
www.iospress.nl

Series B: Physics and Biophysics – Vol.4

Safety Improvements through Lessons Learned from Operational Experience in Nuclear Research Facilities

edited by

Francis Lambert
Atomic Energy Commission,
Paris, France

and

Yuri Volkov
State Technical University for Nuclear Power Engineering,
Obninsk, Russian Federation

Springer

Proceeding of the NATO Advanced Research Workshop on
Safety Improvements through Lessons Learned from Operational
Experience in Nuclear Research Facilities
Obninsk, Russia
1-3 June 2004

A C.I.P. Catalogue record for this book is available from the Library of Congress.

ISBN-10 1-4020-3887-9 (PB)
ISBN-13 978-1-4020-3887-7 (PB)
ISBN-10 1-4020-3886-0 (HB)
ISBN-13 978-1-4020-3886-0 (HB)
ISBN-10 1-4020-3888-7 (e-book)
ISBN-13 978-1-4020-3888-4 (e-book)

Published by Springer,
P.O. Box 17, 3300 AA Dordrecht, The Netherlands.

www.springer.com

Printed on acid-free paper

Printed in the Netherlands.

TABLE OF CONTENTS

Preface

When operating (or dismantling) nuclear facilities, as well as many other complex technological systems encompassing risks, it is necessary to draw lessons from operational experience, in order to improve continuously, and durably improve efficiency and safety.

For that purpose, it is essential to analyse not only the possible significant incidents or difficulties, but also the more frequent mishaps or hitches, or apparently un-significant deviations, in order to early detect any weakness or deficiency in the equipment or organization which might affect safety, and to compensate for it as soon as possible. This is what we call : "lessons learnt from operational experience". (In French: le retour d'expérience des évènements et incidents d'exploitation).

This is important because such analysis allows to consider the potential consequences due to additional deficiencies or less favourable circumstances, and thus to anticipate relevant actions against such consequences (precursor detection).

Another important matter is the necessity to share, between operators of different facilities, about lessons learnt as well as best practices, first inside the same institution, and then on a larger scale.

The main objective of the workshop was to exchange about the organization, that the participating institutions have built, to best draw and implement the lessons learnt from operational experience (current events and minor incidents), in order to improve the nuclear safety of their facilities: research reactors, laboratories and waste treatment facilities. To give an idea, the current events and incidents include most of those rated below INES level 2 (INES : International Nuclear Events Scale).

Besides the incidents, it was intended to also exchange information and experience about the organisation set up to share about operational best practices in routine and non routine operations (such as refurbishment operations).

The means to efficiently gather this operational experience, to record it in order to make it available in the short and in the long term (for the future generations of operators), to share it between the operators of different facilities on the same site and on different sites, were interesting questions for discussion between the participants.

Among the numerous subjects, which deserve to be tackled in that field, are the following :

- tools used for collecting, analysing, reporting, recording, and saving (archiving) all the details necessary to a complete analysis of these events,
- structure of an incidents data base,
- guides for safety analysis of these events,
- recommendations resulting from a statistical analysis of collections of operational events in similar facilities,
- quality system associated with the organization and the management of this field of activities,
- criteria set by the regulator, or the operator, to notify the incidents,
- summary of the reports to the authorities (institutional or regulatory),
- different interpretations of the INES scale, etc...

Out of scope of the workshop however, was the organization to control emergency situations, because this covers more serious incidents or accidents.

The NATO Advanced Research Workshop n° 980580, whose proceedings are presented in this book aimed at bringing together nuclear operators from a group of Eastern and Western countries, selected among the nuclear research and development institutions, rather than power plant and nuclear industry operators, in order to focus the presentations on comparable R and D situations. We are deeply convinced that this workshop achieved its objectives and permitted fruitful discussions upon "how to use the lessons learnt from small events or near misses occurring in the daily operation of nuclear research facilities, to prevent more serious events and thus strengthen the safety", and it was a shared feeling between the participants that these relevant presentations and useful discussions facilitated the establishment of links for future collaboration between colleagues from all these countries.

Francis LAMBERT and Yuri VOLKOV

ACKNOWLEDGMENTS
NATO ARW 980568

I wish to thank first Pr. Yuri VOLKOV from the State Technical University for Nuclear Power Engineering in Obninsk, for chairing with me the ARW dedicated to " Safety Improvements Through Lessons Learned from Operational Experience in Nuclear Research Facilities", and for the excellent collaboration we had during the preparation and realisation of this workshop. My thanks are also directed to Pr. Nikolai SAKOLNIKOV, President of Obninsk Institute for Nuclear Power Engineering, for his hospitality and the warm welcome given to the workshop.

Then I wish to express my admiration and my gratitude for the outstanding work done by Irina VOROBIEVA and her associates, Lena ZINOVIEVA and Polina GERASKINA, and students, for the local organisation of the ARW. This organisation was more than perfect, and every participant did appreciate the availability and kindness of Irina and her team, and the help they kindly provided to each of us upon any request : they did a marvellous job before and during the meeting.

I wish to mention here the efficient work done by the organisation committee: Christine FELTIN, Pascal DEBOODT and Irina VOROBIEVA together with the two co-directors, especially during the preparation phase and the Paris meeting in January 2004.

I wouldn't forget to extend my deep appreciation of the committed help given by Claudine ANGHINOLFI to the production of this publication : I wish to thank her on behalf of all the contributors to this book.

Finally I wish to thank NATO and the Advisory Panel for Security-Related Civil Science and Technology Programme Director : Fernando CARVALHO RODRIGUES, SCK-CEN and the MOL Centre Director : Paul GOVAERTS, CEA and the General and Nuclear Inspection Director : Michel LAVÉRIE for their financial support and their interest, which made possible the realization of this ARW.

Francis LAMBERT

HOW TO EXTRACT THE HIDDEN LESSONS FROM A SMALL INCIDENT ON A NUCLEAR RESEARCH FACILITY

Practices of the French Atomic Energy Commission

Michel Lavérie
Atomic Energy Commission (CEA) France

Abstract: As the operator of approximately 70 highly diverse nuclear facilities, the CEA attaches particular importance to experience feedback on minor incidents: indeed, as safety demonstrations are generally based on the presence of several independent "lines of defence", only through attentive investigation of every occurrence, usually minor and of no consequence, can the level of trust placed in each of these defensive lines be confirmed, or the potential risks arising out of a possible weakness in the system be anticipated.

The efficiency of the system is based on an exacting procedure: Stringent identification of all incidents, consideration of the potential consequences of the incidents in their most pessimistic scenarios, and promotion of a broad conception of transpositions of the events, in time and space, for experience feedback.

This efficiency presumes motivation on the part of all those involved, hence the importance of dissociating from the concept of an "incident" any notion of "error" or "blame" both in internal analysis and in public communications.

The nuclear industry has developed some very progressive tools for experience feedback, compared to practices as regards other technological risks.

F. Lambert and Y. Volkov (eds.), Safety Improvements through Lessons Learned from Operational Experience in Nuclear Research Facilities, 1–21.

1. THE FRENCH ATOMIC ENERGY COMMISSION, AN OPERATOR OF NUCLEAR FACILITIES

The French Atomic Energy Commission (CEA) is a public organisation for research, development and innovation in the fields of energy and new technologies for information and health. It must also guarantee the sustainability of nuclear techniques as a dissuasive force.

In these various ways, the CEA operates a large fleet of diverse nuclear facilities, which it needs in particular:

- for the development of new technologies or improvement of existing technologies in nuclear energy, throughout the entire nuclear cycle

- to provide support for scientific research (physics, medicine, etc...)

- for expertise (in materials, etc...)

- for the management of its own facilities (treatment of effluents, processing and storage of waste, etc..).

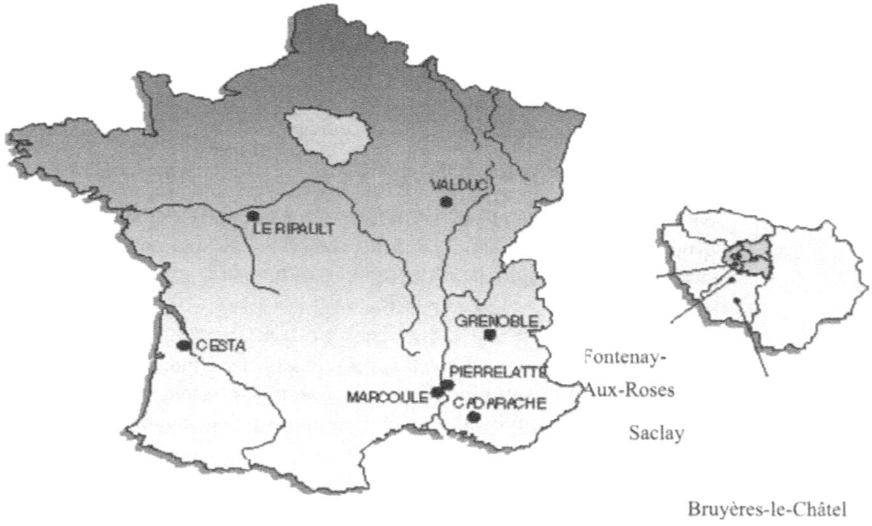

Thus, on 8 of its centres, the CEA operates approximately 70 facilities that are subject to the specific regulations governing "nuclear facilities". Broadly speaking, these facilities can be divided into 4 virtually equal groups:

- research reactors
- laboratories
- effluent treatment and waste processing facilities
- facilities in the process of decontamination and decommissioning

As a reminder, the CEA also operates 250 facilities that are subject to the general regulations governing industrial safety, as a result of the use or presence of radioactive substances.

The CEA's fleet of facilities is characterised by its great diversity: diversity in the nature of the facilities, in the wide range of technologies used, in the age of the various facilities, in the broad sweep of research for which they are used. As an operator, the role of the CEA takes place in a technical and organisational context very different from the circumstances of, say, the operator of nuclear power plants with a fleet of more or less standardised facilities.

In its role as a nuclear operator, safety management is an absolutely vital aspect for the running of research projects and programmes. The CEA favours a progressive approach that places emphasis on prevention and the use of experience feedback to constantly improve the safety of its activity in a totally open and rigorous manner.

One essential means of action lies in optimising its methodologies for the use of experience feedback on incidents. The topic of this workshop is therefore a core priority for the CEA and we are grateful to NATO for organising this opportunity for discussion.

2. WHY WE SHOULD BE INTERESTED IN "MINOR INCIDENTS"

Clearly, a major incident, which in itself results either in significant consequences or a serious threat for the safety of a facility, will require every effort on the part of the nuclear operator.

However, we must also consider that the prevention of a serious accident often depends upon a several-layered set of safeguards, which can be:

- either physical (a typical example would be in a laboratory where safety is materialised by a succession of enclosures)
- or functional (where a safeguard system is triggered in the event of failure of a component or another system)
- or depend upon the operators' running of the facility.

A zero risk situation is impossible to achieve ; therefore the designers and operators use, as a reference for the order of magnitude of the probability[1] of the occurrence of an accident scenario with unacceptable consequences, a very low value (for instance 10^{-7} occurrence per facility per year for one particular set of scenarios).

In a few rare cases, this probabilistic target can be guaranteed by the reliability of a single component (for instance, a reactor vessel, with respect to the risk of sudden breakage).

However, in most cases this very low order of magnitude is impossible to reach with the technology of the components and systems ; in which case the target can be achieved only by having a series of independant lines of defence (each of whose probability of failure is very much higher than the target, but the product of whose independent probabilities enables reaching the target[1]).

[1] This probabilistic presentation is convenient. However, it must be kept in mind that its implementation on non standardised facilities is often unrealistic. It must therefore be able to be transposed into practical modalities of deterministic appreciation for the order of magnitude of the confidence that can be given to each line of defence.

In practical terms, guaranteeing the safety of a nuclear facility is much like assuring the impregnability of a fortified castle.

THE SAFETY OF A NUCLEAR FACILITY :
A FORTIFIED CASTLE

An homogeneous safety, based on several indépendent lines of defence

Let us return, in the general case where safety is guaranteed by *n* lines of defence, to the use of experience feedback on incidents. A "weakness" revealed in one of those *n* lines will in all evidence be a "minor incident": with no resulting damage, and safety being assured by the other defensive measures ...

One might be tempted not to bother with experience feedback in such cases with such low consequences. On the contrary, though, safety must be considered as dependent upon the confirmation, through experience or through the improvements that arise as a result of experience, of the reliability of each individual line of defence. These confirmations and improvements are essentially what assure the progress of safety.

Most serious incidents or accidents are, upon analysis after the event, proven to result from a series of failures, each minor in itself, but some of which were probably less improbable than anticipated.

3. IDENTIFYING AND EXTRACTING INFORMATION ON INCIDENTS

The smooth functioning of experience feedback on minor incidents requires a whole set of stringent practices familiar to and applied by all those involved.

3.1 The notion of "incident" must be accurately defined

It is important, obviously, to have a clear reference about the nature and thresholds of events that constitute an incident. An incident is a deviation from the regulatory reference system applicable to the safety of the installation and any malfunctioning affecting a safety function. In particular, any event that adversely affects the efficiency or reliability of a line of defence is considered as constituting an incident.

Despite all efforts, a definition is not always easy to interpret and therefore a sort of "case law" system has developed between the operator and the safety authority regarding the thresholds that cause an event to be considered an "incident". The safety authority is involved in this system because regulatory practice dictates that all incidents must be declared to the safety authority and a "significant incident report" submitted for each, analysing the incident in detail and explaining what lessons can be learned from the event.

We note that the accepted definition of an incident in France is considerably broader than what constitutes an incident according to the INES scale, and refers to a lower threshold. We describe these additional incidents as being "level 0" or "below the INES scale".

This expansion is important since out of approximately 80 incidents that the CEA declares each year to the Safety Authority, about 75% rate "0" on the scale.

Moreover, the operator deals internally with other, more trivial shortfalls from more detailed non-regulatory reference systems (internal documents and procedures), for which we use the term "anomaly".

However, one has to know where to stop monitoring under these reference systems. In particular, we must be able to follow "best practice guides" which are intentionally ambitious and therefore not always perfectly achieved, without generating unnecessary identification of a superabundance of incidents or anomalies, which would be dissuasive and counter productive.

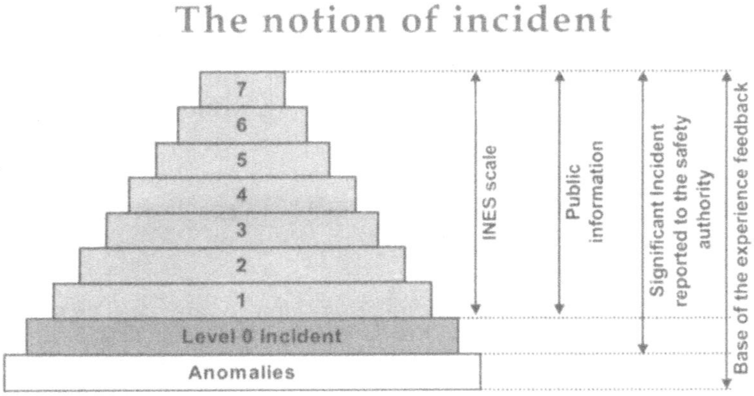

3.2 Responsibility for identification and disposition of an incident

This comes under the province of the facility operator's management hierarchy. The superior in charge must successively:

- declare the incident and propose its classification
- analyse the incident
- define and implement the necessary corrective measures

These different actions are checked and confirmed, at a second level, by a team reporting to the individual directors of centre. The director of the particular centre is responsible for the regulatory compliance of the facility, for all contact with the authorities and all information provided to the public.

3.3 Analysis and disposition of an incident

Beyond informing the CEA hierarchy and the Authorities within 24 hours (mandatory), incidents are analysed and disposed in a three stage process :

Understand what happened

+ Collate and characterise the facts, identify any deviations from the reference system.
  Collation of facts covers all technical, human, organisational and environmental components and characterises them by identifying the faulty statuses and inappropriate human actions taken, compared to the reference system.
  Experience shows that although the technical and environmental aspects are usually accurately reported, a detailed explanation of the human and organisational aspects is more troublesome because of the sensitivity of the persons involved. We will return to this aspect in § 3.5. If investigators come up against a blank wall, a good solution can be for the operating team to work internally with a situational consultant in a context that does not include members of the hierarchy.

+ Construct the Failure Mode and Effect Analysis (FMEA) tree
  Using detailed and carefully crafted techniques, the aim is to seek the causes that triggered the sequence of events leading up to the

incident, and the inappropriate actions taken, in order to identify all of the contributory factors.

- Evaluate the stakes and rank them in order of importance
    + Analyse the real consequences of the incident
      Evaluate any discharges and doses, soil and/or equipment contamination, the impact on activities and installation availability.
    + Analyse the potential consequences of the incident
      Re-examine the consequences, ignoring lucky hazards due to location, favourable statuses and fortuitous human actions and look again to see what the consequences would have been with an identical scenario in a less favourable context (as foreseen by the reference system, without assuming any further failures).
      Evaluate the distance from the feared consequence in terms of remaining reliable and robust lines of defence.

- Decide what corrective or preventive measures to implement
    + Define and implement the action needed on the relevant installation.
      Repair and re-qualification of equipment, decontamination if necessary, identification of the most appropriate corrective and preventive measures, assessment of cost/benefits, preparation, implementation and monitoring of an action plan.
    + Define transverse actions and experience feedback.
      Identification of the generic, transposable lessons learned, and of the other installations concerned. Transposition studies. Action plan to prevent and correct hidden weaknesses.

This approach, briefly outlined above, is discussed in a detailed guide (a CEA internal document) drafted specifically for its operating teams. It should be considered more a guideline for analysis than a rigid framework to straitjacket the thinking of the operating teams.

3.4 Overall analysis of the incidents

In addition to the individual approach to incidents described above, and with the perspective gained over time, it is important to conduct a broader and more synthetic analysis of the characteristics and lessons learned, based on a sufficient number of events.

At the CEA we do this analysis on a sliding 3-year calendar, based on approximately 200 incidents. In this way we can develop a typology of incidents per category of installations, per consequence, per nature, per type of triggering event, per type of cause upstream, and so on.

In illustration of the above, we include below a few examples of histograms illustrating:

- the consequences of the incidents
- the nature of the incidents
- the causes of the incidents

This approach helps us to highlight any recurrent aspects and establish a ranking of efforts.

Diagrams of the consequences

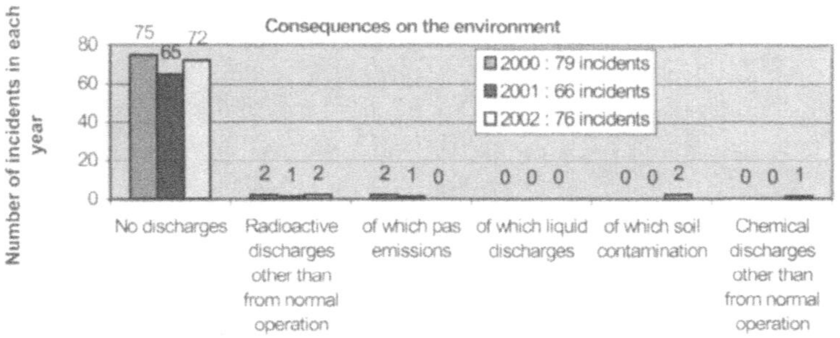

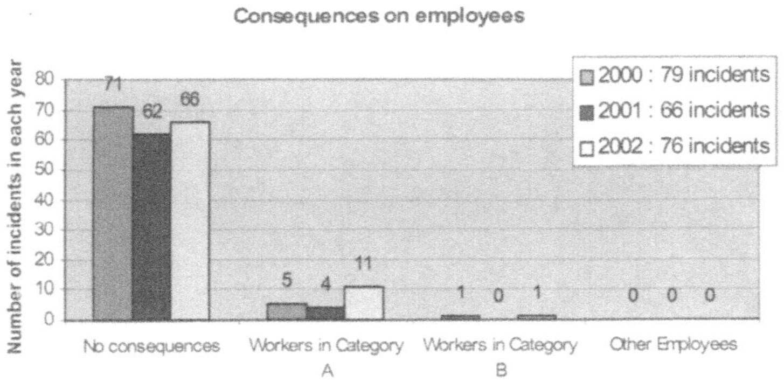

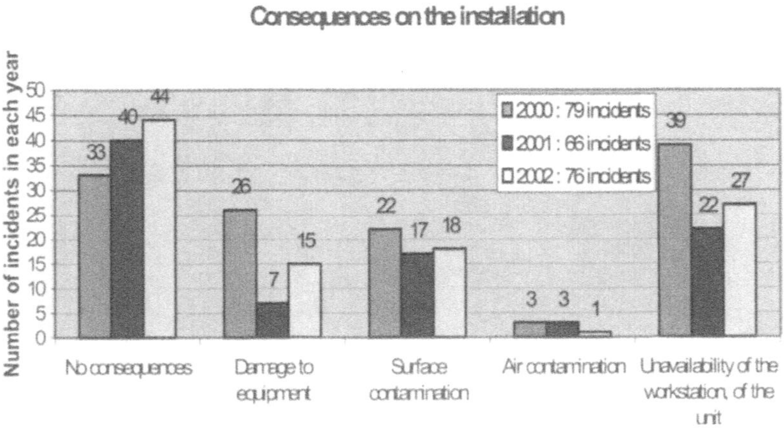

Examples of diagrams per nature of incident, according to the successive levels of detail

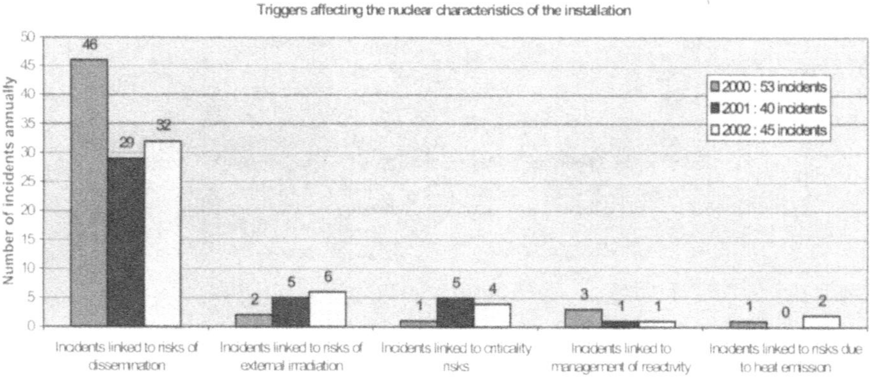

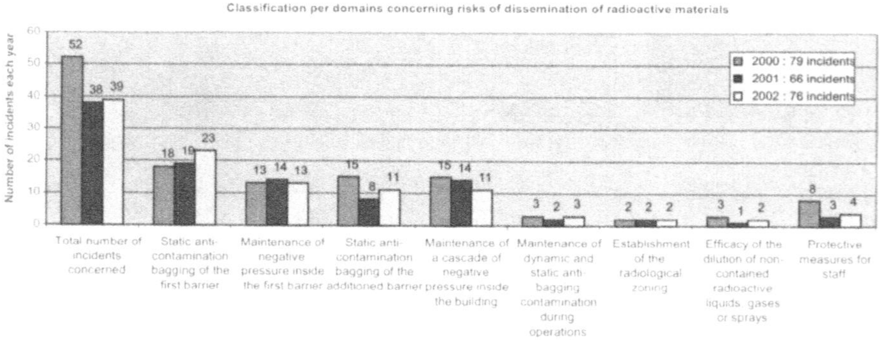

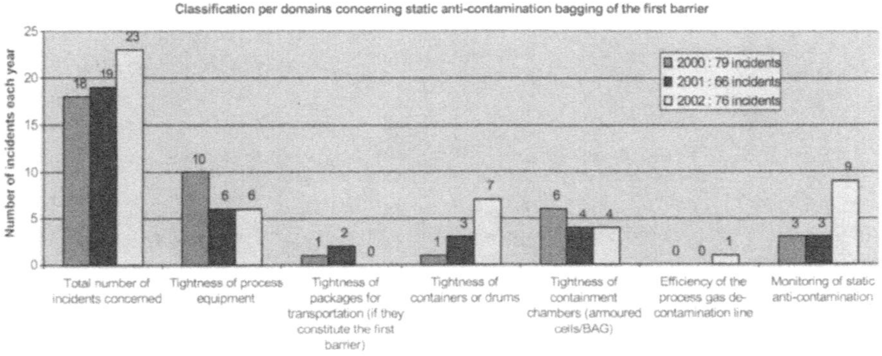
Classification per domains concerning static anti-contamination bagging of the first barrier
Number of incidents each year
25
20
15
10
5
0
2000 : 79 incidents
2001 : 66 incidents
2002 : 76 incidents
18
19
23
10
6
6
1
2
0
1
3
7
6
4
4
0
0
1
3
3
9
Total number of incidents concerned
Tightness of process equipment
Tightness of packages for transportation (if they constitute the first barrier)
Tightness of containers or drums
Tightness of containment chambers (armoured cells/BAG)
Efficiency of the process gas de-contamination line
Monitoring of static anti-contamination

Examples of diagrams per causes

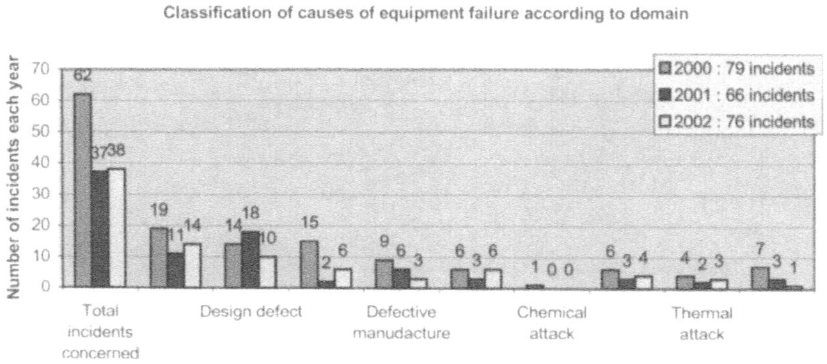

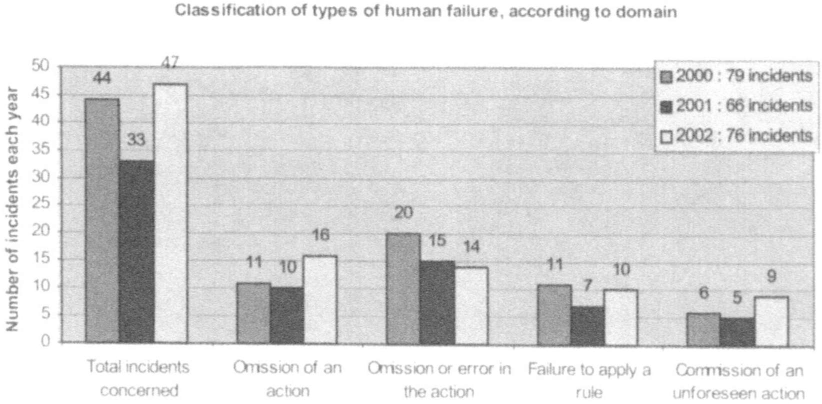

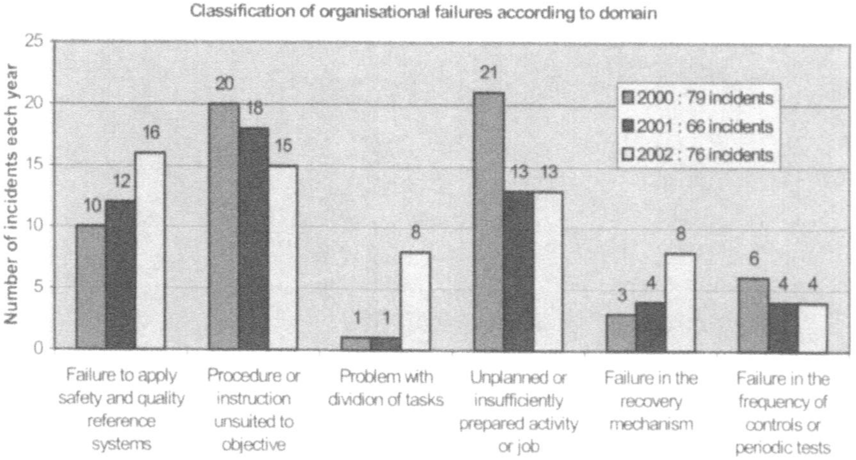

3.5 Peace of mind with the experience feedback system

Using incidents to provide experience feedback is a very powerful tool for the advancement of safety. All players are convinced of its usefulness. However, if the system is to achieve progress, everyone has to participate fully: identifying and treating the causes of incidents requires willingness and team effort, something that no hierarchical or regulatory system of controls can ever replace.

The mechanisms whereby incidents are handled can arouse negative feelings at times : if the safety level is judged inadequate, if a person or a group is found to be at fault or to have exercised poor judgment, if there is a likelihood of sanctions being taken... External repercussions can exacerbate those feelings: if declaration to the regulatory Authority is rewarded by tougher measures, if a press release on incidents ranked on the INES scale unleashes a negative reaction fuelled by the media... Is pointing out incidents therefore not a form of self punishment?

Such perceptions can be detrimental to the performance of the technical tool of experience feedback and can even seriously hinder the process of analysis by hiding the facts. It is important therefore that those in charge (internal and external) without fail provide tangible evidence that the identification and handling of incidents are technical measures intended to achieve progress rather than serve as the basis for a system of assessment or apportioning blame or passing judgment on a person or a group of people.

In order not to lay blame on (or perhaps disculpate) the operating teams, the entire process should take place at line one of the system. This result, however, is not always fully understood, and is still a fragility.

4. A FEW METHODOLOGICAL CONSIDERATIONS

In the general approach described in chapter 3, I want to emphasise 3 methodological steps that are necessary if the incident analysis is to be fully effective. Without these steps, the experience feedback would provide nothing more than mere occasional corrections of discovered failures.

As part of the operators' safety culture, we must make every effort to promote these 3 steps. They are sometimes received with reluctance or concern, inasmuch as they may, upon analysis, reveal that a factually benign incident is an important instigating event requiring an action plan of broader scope than initially thought.

4.1 Consider the potential consequences in their most unfavourable circumstances

As said in § 3.4, the aim is to eliminate the lucky hazards of location, the fortuitous statuses and favourable human actions and to re-examine that the consequences would have been if the exact same incident had occurred in the most unfavourable circumstances (as foreseen in the reference system, and without positing any additional failure).

This presumes deep thought to transpose the incident in time and space, asking oneself such questions as:

- On what similar object could the incident just as easily have occurred ?
    - For instance : a sealed container, part of a set of similar containment arrangements (waste drums, glove boxes, etc.) breaks open. We will consider the similar object the contents of which, if released, would be most prejudicial.
    - For instance : 2 objects become swapped around during manipulation. Are there any other swaps in a similar kind of scenario, that could have been more damaging ?

- Might the failed equipment involve a more important source term?
    - For instance : an effluent tank leaks. We consider the maximum source term that it could contain, in accordance with the safety reference documents and the procedures.
    - For instance: a fuel assembly is involved in an accident scenario. We will consider the fuel in the most prejudicial condition in which it could be or have been present (combustion rate, etc.)

- Could the environment of the incident be more vulnerable?
    - For instance : an explosion, steam leak, heavy object dropped in a room where no one is present. If the reference documents authorised the presence of personnel at the time of the incident or at any other time when the incident could just as well have happened, we will consider the potential impact on that employee.

4.2 Evaluate the solidity of the defence in terms of residual depth

A nuclear operator who has a highly detailed probabilistic study on his installations (operator of a standardised fleet of nuclear power plants) can present the approach as follows: an accident scenario has a known probability of occurrence in the nominal safety condition of the installation: what becomes of that probability in a situation degraded by the incident situation occurring?

An operator such as the CEA, who cannot have that kind of probabilistic studies for its many, diversified installations, must still transpose the incidents in the same spirit, using a deterministic approach: in the normal situation, there are *n* lines of defence to counter a feared accident scenario. To each of these lines is assigned a coefficient of confidence (on a sliding scale from "strong line" to "weak line") that is used to weight the addition. During the period in degraded condition, this same count can be made about the remaining lines.

Schematically, we can describe the procedure for analysing an incident by saying that:

- in the incident identification phase, we will above all investigate the line (or lines) of defence that were weakened or lost;
- in the evaluation and ranking of stakes phase, the interest will shift to the remaining lines of defence.

Whenever the analysis reveals a particular weakness or "thin" spot in the remaining lines of defence, we can qualify the incident as a "precursor" and mobilise a great deal of effort on remedying the situation.

4.3 Define the field concerned by the experience feedback

It is important that thinking on experience feedback should spread outwards from the incident in a succession of widening circles:

Stage One: analyse the failures and define "spot" corrections to avoid a recurrence of the identical incident.

Stage Two: analyse the field that presents a similar vulnerability in the face of an identical scenario and define a broader set of investigations and corrective measures on the affected installation and on other installations.

Stage Three (where necessary) : after gleaning a thorough understanding of the upstream causes of the incident and highlighting a failure "context" (technical shortcoming, poor organisation, etc.) in answer to the question "how did this happen", we should then ponder the full extent of the field that could have been impacted by a similar context.

THE FIELD CONCERNED BY THE EXPERIENCE FEEDBACK

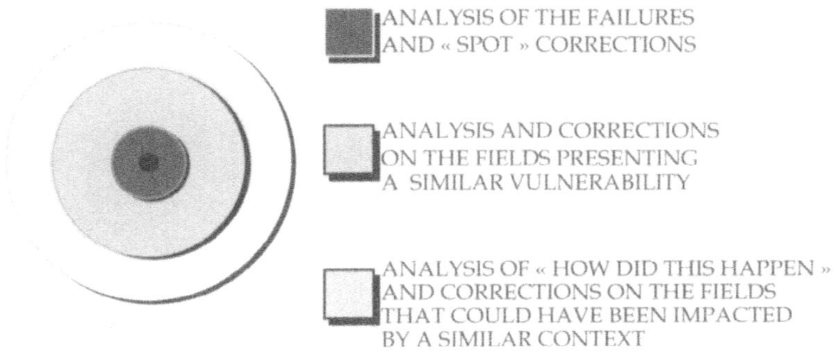

4.4 A simplified example of application of these 3 methodological steps

This is a simplified example based on a real incident that occurred at the CEA (the only incident ranked in category 2 of the INES scale in recent years).

A package of material B, brought into a laboratory booth, proved to contain less nuclear material than indicated in the accounting inventory. The cause was quickly determined: after a problem with marking, package B was mistakenly substituted for a package A which was to be used, in the nuclear materials shop.

We found:

- that the real consequences were minimal, other than a disturbance of the laboratory work schedule;
- that, on the contrary, the concept developed in point 4.1 raises important questions: could a package become swapped for another in no less likely circumstances, that would be more prejudicial, particularly with regard to the risk of criticality in the booth?

- that the concept developed in 4.2 led to the evaluation of a set of significantly weakened lines of defence.

Consequently, decisions relative to the field concerned by the experience feedback have been very tough: i.e. to take back and check every inventory (location, identification, composition) of all the items comprising nuclear materials in the whole CEA.

5. CONCLUSIONS

Without an active and demanding incident analysis (including minor incidents) and experience feedback system, nuclear safety would be imperilled. It would lack two basic ingredients:
- Confirmation through operating experience of the robustness of the lines of defence making up that safety, and the consolidation of potential weak points;
- Detection of shortcomings that can occur gradually over time (aging of components, deteriorations due to the circumstances of operational organisation, etc.).

Of all activities comprising technological risks, the nuclear industry appears at the current time to be exemplary in terms of experience feedback on incidents.

This exemplarity is based in particular on the formal recording and traceability of the disposition of incidents, down to a very low threshold of importance :
- broad definition of the incidents to be investigated and procedures for their disposition prepared by the operators ;
- declaration of all incidents that occur and a report sent to the Regulatory Authority on how they were handled ;
- information of the public regarding all incidents that fall into any category of the INES scale.

However, there is a downside to everything: because the nuclear industry is more demanding in its standards of reporting and communicating, we must be careful not to convey an untrue image of the activity as encountering more preoccupying events than do other industries.

We should note that following a major accident of a chemical nature that occurred in France near Toulouse in 2001, the Parliamentary Commission of Enquiry recommended that industries involved in at-risk technologies model their practices more closely on those of the nuclear industry.

To conclude, I want to emphasise the importance of the team spirit that must exist in any operating organisation if incidents are to be handled effectively. Attachment to the common goal is not something that we can count on as a due, and various considerations may generate reluctance:

- the handling of incidents requires resources;
- it ruffles feathers, by pointing the finger at malfunctioning that calls into question the choices made, the organisation, the people;
- lastly, it sometimes leads to complicated and costly preventive or corrective actions.

The handling of incidents therefore derives from a strong culture of safety. The entire hierarchy must engage in this area, willingly and with determination, from the top down, in a real effort to develop a culture in the operating teams that will enable the identification and feedback of all the necessary information.

However, nothing in this area can ever be considered definitively won.
This workshop, for which I thank the organisers, affords us the opportunity to gather together and discuss our mutual experiences, and is thus an excellent forum for progress.

LEARNING FROM LOW LEVEL INCIDENTS

A revised approach to the recognition, reporting and analysis of minor incidents and mechanisms adopted to cascade the lessons learned

Alisdair R. Burnett
UKAEA Dounreay

Abstract: This workshop is dedicated to learning from low level incidents, which taken together indicate areas for significant loss, if no appropriately acted on. These low level "precursors" in themselves do not result in any measurable detriment to the organization, unlike the significant incidents which normally are the ones given most attention and follow up.

This paper is configured around a new approach to incident reporting, investigation and learning of lessons at UKAEA Dounreay. This approach has two clear objectives:

- to improve the numbers of low level incidents reported by encouraging all employees to report near misses. To assist in this, a range of options for reporting were put in place and a new, more versatile data basewas commissioned to capture and analyse the information;

- to expand on the sources of lessons learned from incidents and to raise the effectiveness of the cascade mechanism for these lessons, measuring improvements by monitoring subsequent incident trends.

F. Lambert and Y. Volkov (eds.), Safety Improvements through Lessons Learned from Operational Experience in Nuclear Research Facilities, 23–46.

1. INTRODUCTION

As a backdrop to the presentation, I will start by giving a condensed overview of the site plant facilities – their history and current operational status.

The Dounreay site comprises of a mix of experimental fast reactor prototypes and their associated fuel reprocessing facilities and employs around 2000 staff (60% permanent staff and 40% contractors). The reactors, two fast breeder reactors and a materials testing reactor (MTR), utilised sodium (Na) or a mixture of sodium and potassium (NaK) as the coolant medium. The first reactor to go critical (in 1959) was the Dounreay Fast Reactor (DFR) and it operated at a maximum 60 MW thermal (15 MW electrical) and played an important part in the world's fast reactor research programme by developing fuel designs, coolant technology, efficient reprocessing and nuclear waste management until its closure in 1977.

Slides 2 - 5

Learning from Low level Incidents
Materials Testing Reactor
UKAEA

Learning from Low level Incidents
Dounreay Fast Reactor
UKAEA

Learning from Low level Incidents
Prototype Fast Reactor
UKAEA

Learning from Low level Incidents
UKAEA

This was followed by the Prototype Fast Reactor (PFR) which had an electrical output of 150 MW and operated from 1974 to 1994. As well as operating as a power station it provided information for the future design and operation of large commercial fast reactor stations. Following shutdown of the reactor, Stage 1 decommissioning started and is expected to be complete in about 2009. All fuel has been removed from PFR and the next phase of work is to treat the 1500 tonnes of sodium used as a coolant. A £17 million Sodium Disposal Plant has been built in the former turbine hall to remove radioactivity from the sodium and convert the rest of the material to a salt that can be discharged safely to sea.

The reprocessing plants comprise of a suite of facilities which have the capability to manufacture and reprocess a range of fuels, for example our MTR fuel element plant (D1202) produced elements and fuel plates for three research material testing reactors and an adjacent processing plant (D1203) was used to produce Uranium billets from uranyl nitrate, a product from two associated irradiated fuel reprocessing plants.

These plants are supported by a range of waste storage facilities.

2. CONTEXT

This Paper is configured around a new approach to incident * reporting, investigation and learning of lessons implemented at Dounreay, primarily triggered by 2 high profile accidents which both occurred in 1998.

At this juncture I will define the use of the term 'incident' as used at Dounreay. An incident includes both potential or actual loss to the organisation in terms of injury, plant damage or harm to the environment. The term is also sub-divided into Unusual Occurrence(**UNORs**) which encompass low level or 'near miss' minor loss happenings (or precursors) and **Events / High Potential** which are incidents involving significant actual of potential loss.

Slide 6

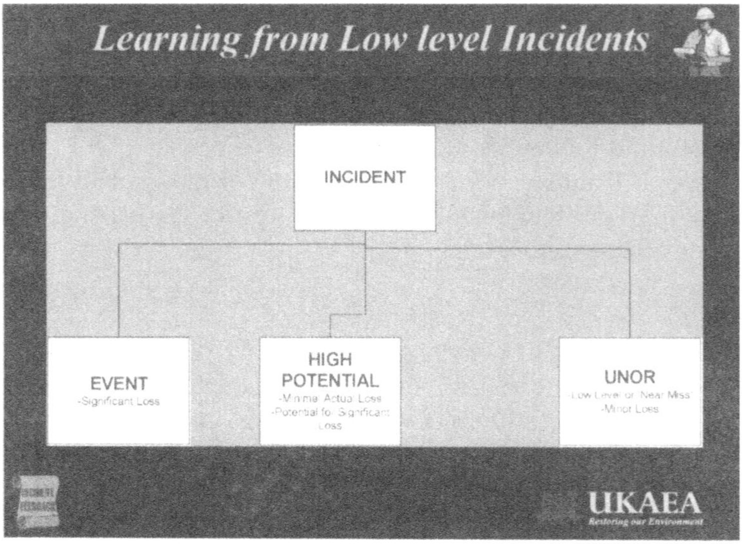

On a timeline through the mid to late nineteen nineties, our UNOR / Event reporting statistics were manifesting themselves as follows :

Slide 7

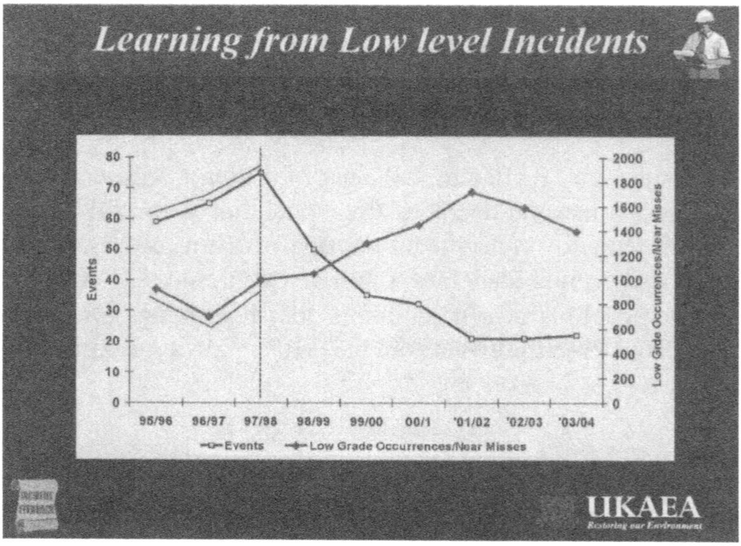

which indicate a high proportion of Events to low level precursors i.e. 75 : 1000 (1 :13).

The two incidents which occurred at this time were both non-nuclear in nature, although one which involved damage to a high energy 11 kv electrical supply cable shutdown our entire reprocessing facilities and had major repercussions over the subsequent 4 years. This incident resulted from a mechanical digger contacting the cable and shorting out the supply, leaving the plants without space extract ventilation.

The cable incident was accorded Level 2 on the INES scale and resulted in a major Regulatory Audit of all our arrangements and operations.

The other incident was a severe industrial injury to a delivery driver, who was unloading 200 litre empty waste drums from the back of a lorry and was struck on the neck by a drum, which left him paralysed from the neck down.

The 2 critical aspects of the incident investigation processes which were targeted by these Events were :

A) Initial reporting mechanisms – linked to awareness and understanding of what and when to report ;

B) The need to have effective 'lessons learned' capability through the exchange of operational lessons throughout industry – and feedback processes to staff which are effective.

3. INITIAL REPORTING

The critical aspect A related to the drum handling injury, whereby previous to the incident, there were occasions where drums being unloaded narrowly missed the driver but as no injury occurred, it was erroneously interpreted by those witnessing these 'near misses' that nothing need be done and the observed unsafe practices went unreported. i.e. there was no loss incurred but the potential loss aspects were not recognised.

A key finding of the formal accident investigation was to ' Improve the recognition and reporting of 'near misses''.

The critical aspect B related to the cable incident; a key recommendation from the final audit was to improve UKAEA's arrangements for obtaining and learning from relevant information on events, incidents and accidents occurring outside the UK.

Starting with A on improving initial reporting of unusual occurrences, a task team was convened and considered the following :

i)

- to open up the options or mechanisms for registering incidents and to make it as easy as possible for those who observe an incident to simply report it.

ii)

- to publicise / promote these options or mechanisms to the whole workforce and to explain crucially why it is being done i.e. what are the benefits.

iii)

- to record / collate the reports generated on a versatile database which will be capable of analysing the data to record emerging trends and patterns. Central to this system is the application of consistent and effective root cause analysis.

Slides 8

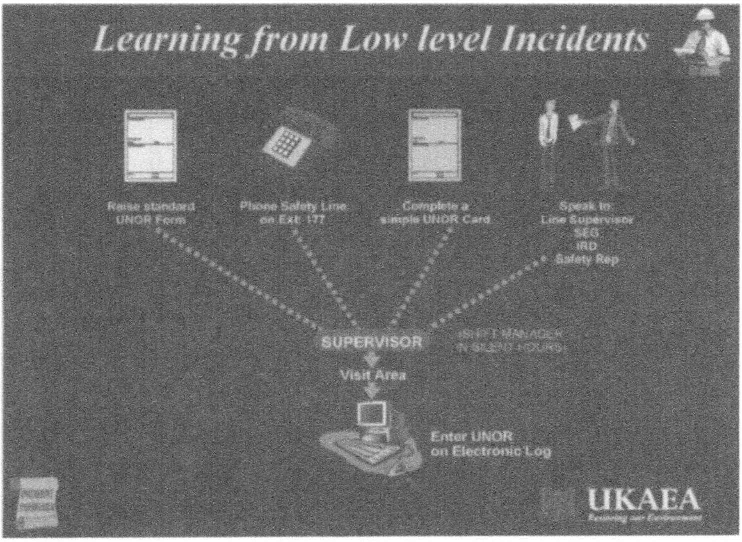

3.1 Options for Reporting

At the time of the incidents, the only existing mechanism for initial reporting of unusual occurrences was a slightly cumbersome report form which had to progress through a number of hands for information and action. This inevitably led to delays in the facts of an incident becoming known to the central safety support team – whose immediate remit is to onward report to Regulators, set the appropriate level of investigation and disseminate the facts as known to responsible managers. Having only one option also has an adverse impact on the overall level of reporting.

In light of these issues, it was decided to open up the options on reporting such that the originator/witness of an unusual occurrence would select a mechanism which best suited themselves. As such 4 options were chosen, each being simple and straightforward, e.g. :

1. phone a safety helpline (ext 177) and give brief details of the incident – what happened, where and when – to the telephone operator. This information can be passed on anonymously – as some people refer not be named;

2. raise a simple yellow 'UNOR' card giving brief details of the incident and post in a box for subsequent collection and processing (throughout the site there are 8 post boxes strategically placed to allow cards to be deposited);

3. speak to a Safety Representative, Safety Advisor or Line Supervisor who will assist you in recording the incident details;

4. raise a standard Report Form electronically yourself via the site intranet.

The essence to this revised approach to the 'front end' reporting was to remove perceived inhibitors to reporting, whether it be reluctance of a witness to give their name if they have observed an unsafe practice or whether they are hesitant in approaching a supervisor of the area to report the occurrence.

It is known from studies that expanding on the options available for initial reporting of unusual occurrences and including the option of anonymity, does improve on the number of 'unsafe practices' recorded. This is often due to the reluctance of a witness to report on a co-worker or colleague who is doing something unsafe.

Of equal importance is the need to give feedback to those who report unusual occurrences such that they are informed of progress and what follow up action is being planned.

3.2 Publicising and promoting the reporting mechanisms

Concerning publicising and promoting these reporting mechanisms to the site workforce, a concerted advertising campaign was launched which included a progressive communications pitch involving :

- Safety Information Roadshows (information stands with exhibits, demonstrations, advice etc)- sited in local plant areas;

- Safety Information Briefs - articles delivered by supervisors to team Members;
- Presentations by Operational - key points delivered to front-line supervisors;
- Feedback Section and plant managers.

3.3 Capturing the information on a versatile database

Slides 9,10,11

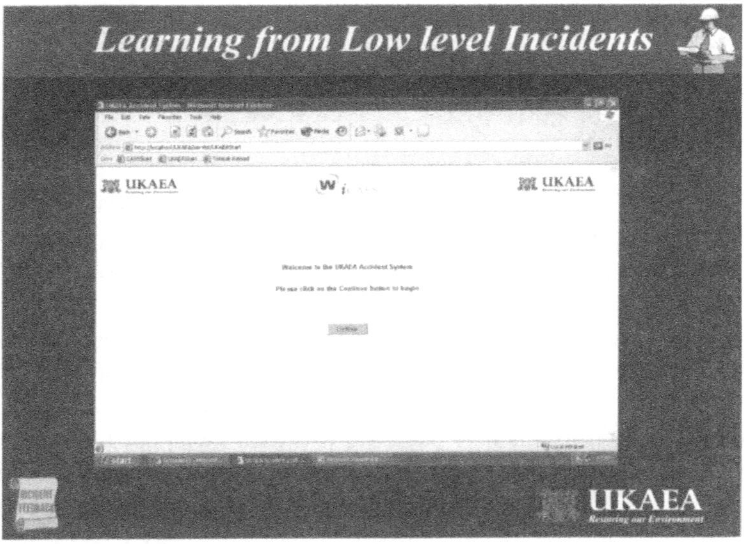

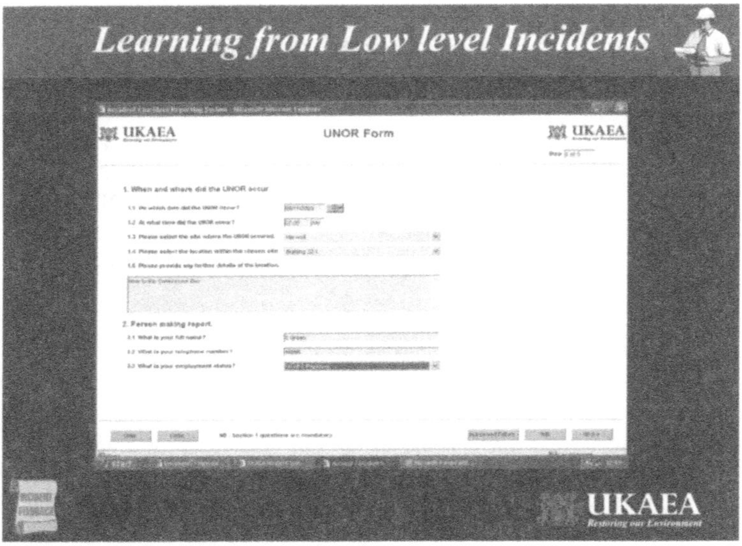

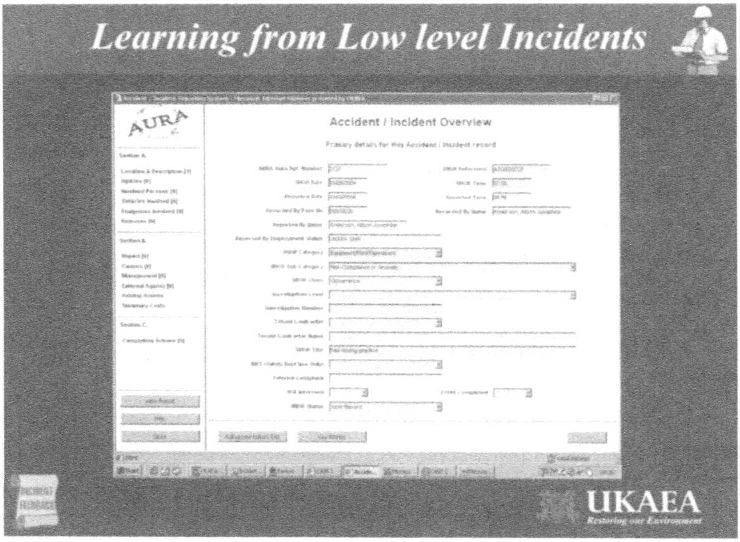

In the months following the incidents, a new database was introduced to address 'front end' reporting and facilitate prompt dissemination of the reports to responsible managers – in the first instance, a new UNOR record is directed to the local area supervisor for initial action. The supervisor entered this information (originating from any of the 4 options mentioned) into the database and cascaded it to other interested parties throughout site.

The information from this new database was then transposed into an existing incident analysis database to produce trending reports , causal analysis etc. Latterly, in late 2003, these databases were merged into a single composite inputting and analysis database, termed AURA (Authority UNOR Reporting & Analysis) which has recently been fully commissioned and has been in operation for six months.

AURA's main attributes are as follows :

a) it has an 'Initiator' page on the intranet, such that any person on site can access this to register a safety concern by logging some simple details - and no prior training is necessary.

b) these Initiators are directed to relevant area supervisors / plant managers for upgrade and action e.g. investigation/causal analysis and action close out.

c) a suite of customised analysis reports can be accessed by plant managers to generate appropriate trending reports to be tabled at plant safety performance meetings and actioned accordingly.

These analyses can be configured at :

- Plant level
- Department level
- Site level

In this respect, for example, an analysis report could be generated at site level if a trend is emerging on lifting deficiencies for nuclear cranes, which may involve a design fault common to a particular type of crane, in use in a number of facilities throughout the site. Also, for a particular plant, there may be a developing trend of head injuries due to a particular work programme (wearing respirators during restricted area working resulting in head knocks from limited vision) which is not an issue at Department or Site level.

4. PREVENTATIVE ACTIONS FOR HIGH POTENTIAL /NEAR MISSES, CORRECTIVE ACTIONS FOR LOSS INCURRING EVENTS

The aim throughout is to have a quick and easy means of reporting so that low level 'precursors' are being identified and reported, resulting in them being fed into the AURA system, for subsequent follow up and trending or pattern analysis. This then allows management to be atuned to problems and can take appropriate early action. linked to this improved reporting is to have staff at supervisory and plant management level properly trained in investigative techniques such that the underlying causes of incidents are exposed and appropriate actions put in place.

In this respect a major training campaign was undertaken to turn key staff into competent investigators. With this in place, causal analysis from plants and departments could be fed back to the central safety feedback unit so that behavioural or job system factors which are sub standard can be assimilated and highlighted to senior managers for remedial action.

As an example of these behavioural and job system factors and concentrating on **no loss incidents** involving high potential near misses, analysis reports on these incidents over 6 monthly intervals have identified common underlying causes of :

- Poor communications between work groups
- Training needs analysis
- Inadequate work standards relating to inadequate monitoring of compliance
- Inadequate hazard identification and risk assessment

Slides 12, 13

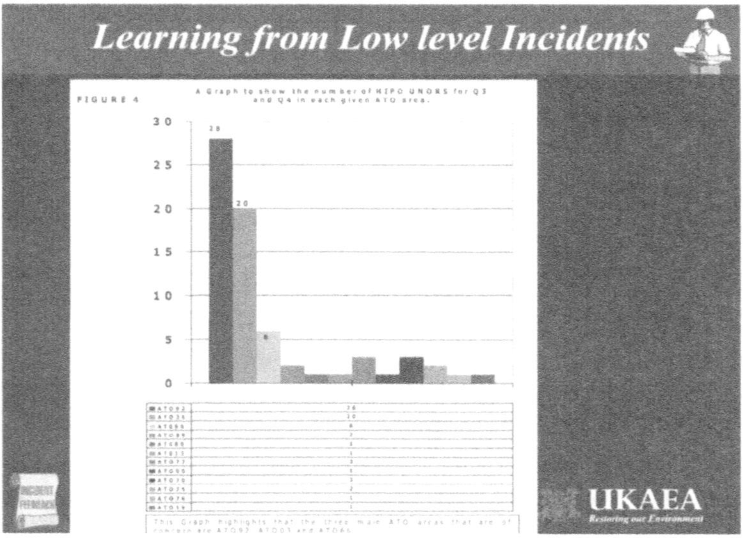

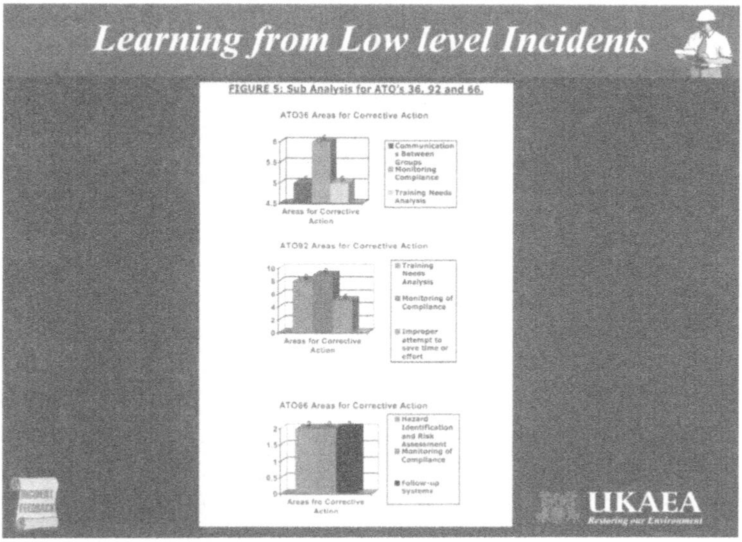

This is a powerful means of demonstrating that carrying out root cause analysis on a large enough pool of low consequence incidents can indicate areas of significant potential loss and also target appropriate management preventative action.

An interesting finding of the most recent 6 monthly analysis has shown that the majority of the immediate causes (unsafe practices committed by employees) are due to there being predominantly job system failures (underlying causes) at work.

Slide 14

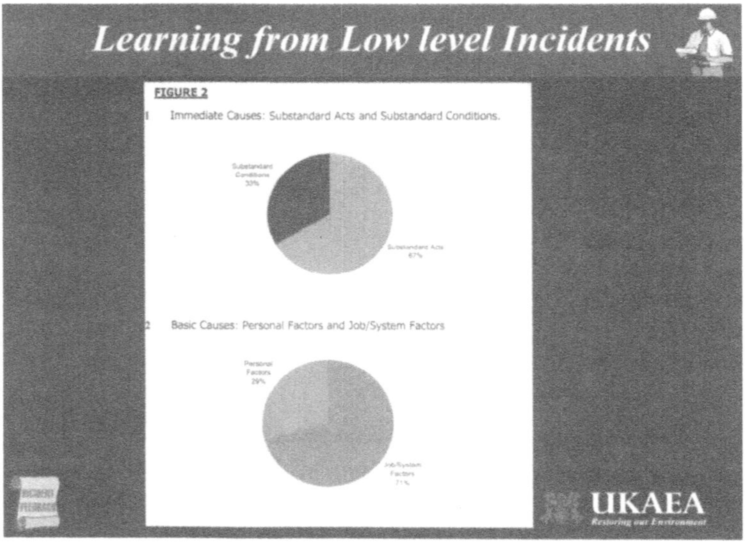

With regard to how we take forward corrective actions from these analyses, we have nominated 'champions' in posts who have specific remits for dealing with the causal factors identified in the analyses, e.g. communications, training needs, monitoring compliance and risk assessment. The way forward is that these champions liaise with the senior managers heading up Operations and Decommissioning Departments with the aim of translating the findings/shortfalls into workable actions.

These types of analyses play a part in the overall trending and analysis of incidents and on a regular (monthly) basis, incidents are discussed at a Loss Control Group (Incident Panel) to identify emerging issues from low level incidents, such that Management can take appropriate action.

On a wider scale and as a means of assessing implementation on best practice for incident reporting, investigation and analysis, UKAEA are a member of the International Safety Rating System (ISRS) which is a monitoring tool used to assess safety performance against set criteria and involves regular audits carried out to determine the compliance level with

ISRS. Currently we are at Level 8 on the ISRS system, Level 10 being the maximum achievable level.

Slide 15

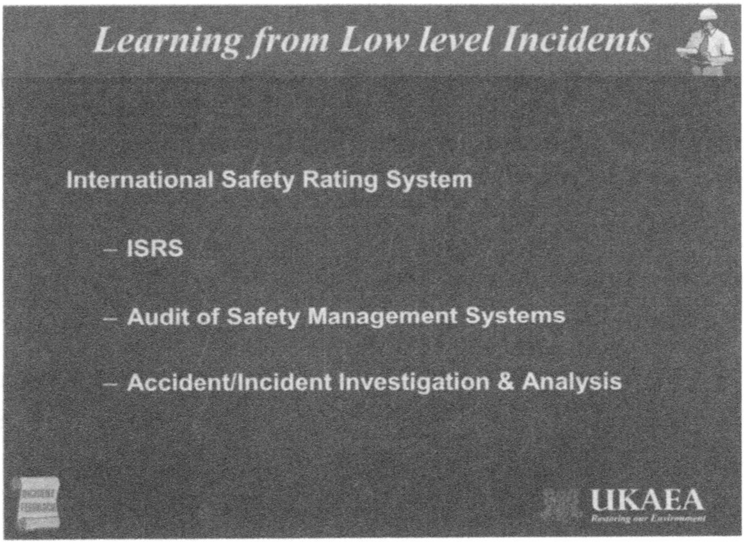

5. FEEDBACK MECHANISMS

With reference to the B critical aspect of the incident investigation process : ' The need to have effective 'Lessons Learned' capability through the exchange of operational lessons generated throughout industry', sources of lessons learned were trawled to identify those which would be suitable for our current business of decommissioning a major nuclear site. A major source of incident lessons was identified at international level and this comprised of the US Department of Energy (USDoE) which regularly publicises short summaries of key incident lessons involving a wide spectrum of construction, decommissioning, radiological, electrical, lifting etc incidents. Further sources identified included, within the UK, other nuclear operators e.g. British Energy, BNFL.

Once the source of lessons had been identified and assimilated, effective feedback mechanisms to staff were looked at. A range of informal processes, mainly paper based, existed which involved disseminating reports or one page summaries to plant managers, such that they would digest and pass on the salient points to their staff.

In order to improve the exchange of information and lessons via direct engagement with the operators doing the work, an 'Incident Feedback Forum' was formed whereby 3 to 4 high profile contemporary incidents (drawn from on site and external e.g. USDoE) are selected and the key points delivered by a series of bullet points which comprise of :

- What happened (the incident)
- Why it happened (root causes)
- Lessons to learn

These key points are supported by photographs and diagrams for emphasis and very recently we have added costs to incidents to bring home the losses incurred.

Slide 16

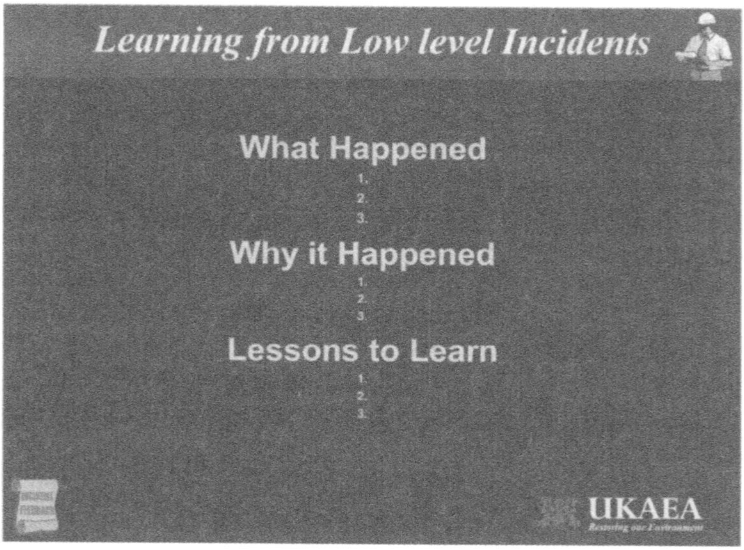

This presentation is implemented by staging local plant sessions with the operators and the delivery is by a Senior Manager. The process has been developed and improved on by consulting with both managers and safety representatives. It is staged in 4 separate venues across site and one of the sessions is filmed so that an archive record is kept of the proceedings for future referral. It also enables those who were unable to attend the sessions at the time to have an opportunity to subsequently view the presentation. This is particularly of benefit to shift workers.

Latterly, we have been looking at supplementing these formal sessions with lessons linked directly to work packages or projects such that during the planning phase of a job, a search is carried out of any known relevant past incident lessons e.g. for a confined space or heavy lifting task and the information (usually in the form of a one page summary with illustrations) is added to the work package.

Slide 17

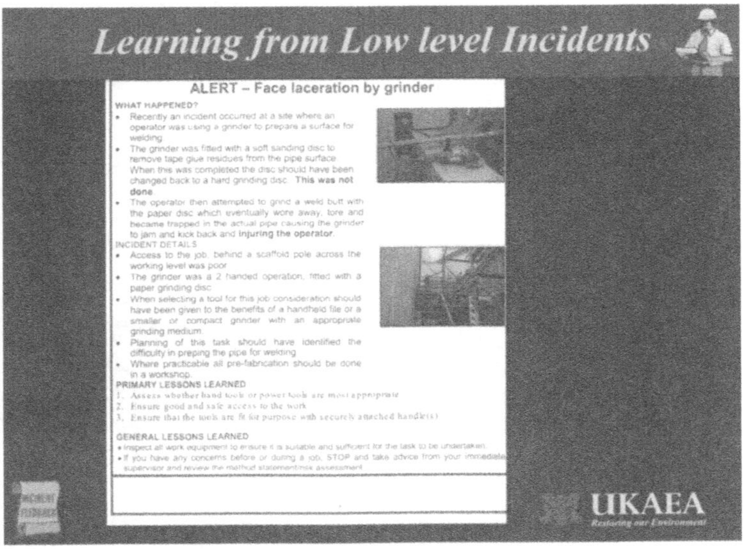

The process in practice would be that the Works Control team who draw together and sanction packages of work would include the relevant learning lesson in the planning stage and it would be disseminated at the pre-job briefing to the work team.

6. CONCLUSIONS

6.1 Results

A clear aim of the initiative to improve the awareness throughout site for all employees to recognise the need to report unsafe acts and unsafe conditions has had a degree of success when a comparison is made with our UNOR reporting statistics over the last ten years. This is best illustrated by the following graph -

Slide 18

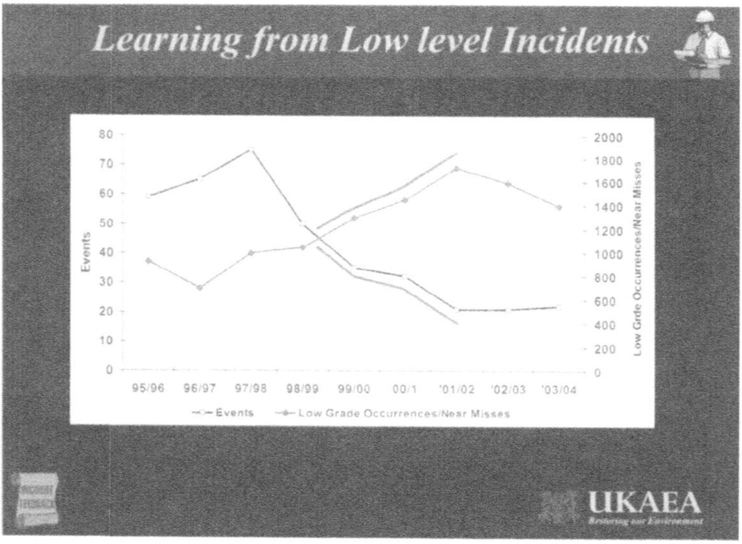

This depicts our site performance over roughly ten years with regard to the numbers of significant UNORs (Events) plotted in parallel with the numbers of low level UNORs recorded. The vertical axis on the left depicts Event numbers ranging from around 80 down to 20. The scale on the right gives low level UNORs or 'precursors' which go from 500 up to 1750.

Quite clearly it shows a progressive improvement in the numbers of low level UNORs reported (from 650 in 1996 to a high of 1750 in early 2002) with a corresponding decrease in Events occurring (75 in 1998 dropping to 20 in 2003) which indicates that trending and pattern analysis for generic underlying causes (from increased reporting) has had a measureable effect. Of particular note is the step change in reporting at around 1998/99, when the improved reporting initiatives came into place (ratio is now 1: 87 compared to 1:13 in 1998).

With reference to an industry wide safety performance parameter of rate of lost time accidents accrued by an organisation, the following graph gives a similarly encouraging picture.

Slide 19

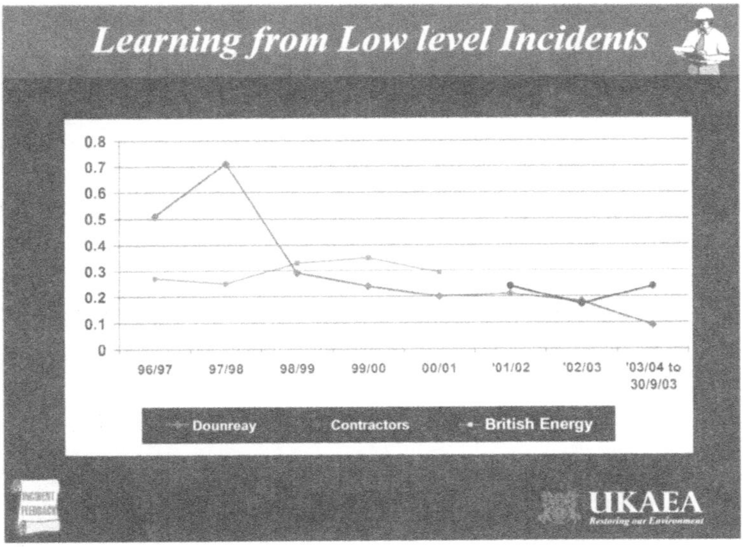

It shows that for the same ten year period, the Accident Frequency Rate (AFR) has dropped from a high of 0.7 to an industry bench mark value of ~ 0.2, indicating that harm to people incidents have fallen considerably.

The AFR (one day or more lost time accident) normalises the number of hours worked and the number of staff employed to give a standardised accident frequency rate across industry which allows comparisons to be made on safety performance.

Very recently, however (last 18 months), there has been a dip in the level of minor UNORs reported (1750 down to1500) and this has been accompanied by an increase in Events. We are currently reviewing possible underlying issues which may be contributing to this recent trend and one particular aspect under consideration is 'Repeat Events' and their frequency in relation to the time taken to close out actions from investigations. Currently repeat events are occurring, on average, every two months whilst investigation actions are taking 8 months to complete. In light of this, we are introducing a new standard on the investigation process such that actions raised which directly relate to preventing a recurrence are closed out within 1 month.

Slide 20

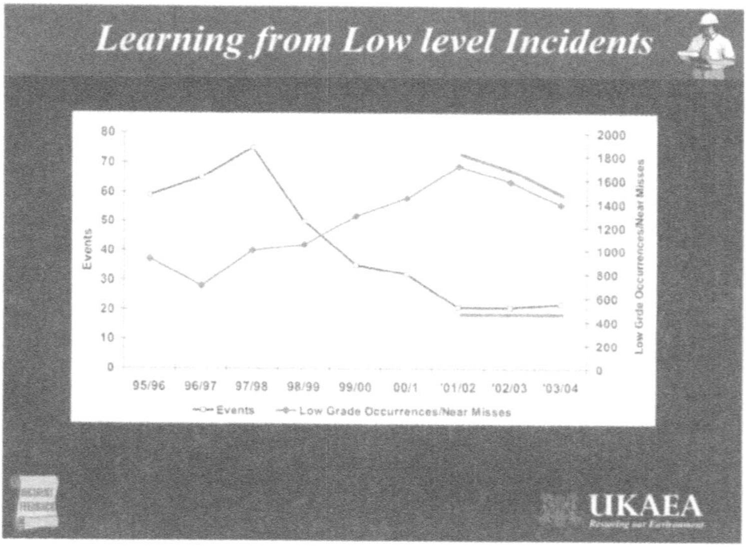

6.2 Lessons learned

Concerning the cascading of lessons learned, we have seen some improved benefits from both the identification and release to the workforce of wider lessons from industry and the formal Incident Feedback sessions. However, with a rapidly changing workforce, the need to refresh archive lessons and inject them directly into relevant planned work is considered to be of major benefit and is receiving due priority for implementation in the near future.

Another aspect of loss control which I briefly touched on earlier and can have a notable impact on driving home key lessons, relates to the costing of losses and in the past 6 months our Incident Feedback Section have used a simple algorithym and ready reckoner to cost around 700 UNORs raised. This exercise has shown that the significant incidents have cost the organisation from £30k to £150k and these figures can give a powerful message with regard to the objective of avoiding a recurrence - costings have been incorporated into recent Incident Feedback sessions. Additionally, for the majority low level occurrences, grouping the costs for the main contributors e.g. equipment damage, injury, environmental damage and fire – and aligning them to the main Departments on site, helps to identify where the major losses are occurring and more importantly where actions are best placed for dealing with these losses.

Slide 21

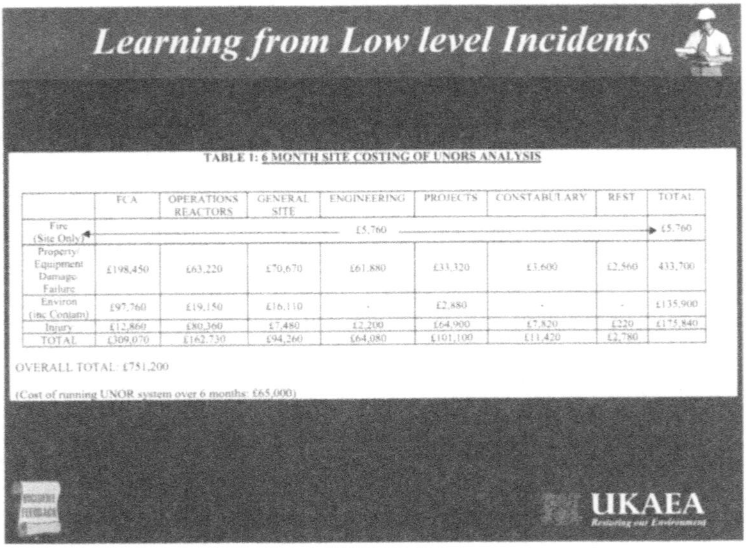

TABLE 1: 6 MONTH SITE COSTING OF UNORS ANALYSIS

	FCA	OPERATIONS REACTORS	GENERAL SITE	ENGINEERING	PROJECTS	CONSTABULARY	REST	TOTAL
Fire (Site Only)	←			£5,760			→	£5,760
Property/ Equipment Damage Failure	£198,450	£63,220	£70,670	£61,880	£33,320	£3,600	£2,560	433,700
Environ (inc Contam)	£97,760	£19,150	£16,110	-	£2,880	-	-	£135,900
Injury	£12,860	£80,360	£7,480	£2,200	£64,900	£7,820	£220	£175,840
TOTAL	£309,070	£162,730	£94,260	£64,080	£101,100	£11,420	£2,780	

OVERALL TOTAL: £751,200

(Cost of running UNOR system over 6 months: £65,000)

7. CLOSING STATEMENT

Clearly benefits have accrued from Dounreay's revised approach to the reporting of unusual occurrences, which allied to competent investigation and dissemination of the key learning lessons, have shown a marked downward trend on the numbers of Events occurring over the last 10 years.

A note of caution is introduced, however, in relation to recent trends whereby the Event / UNOR graphs are starting to converge. In this context efforts are being directed to evaluating and taking early action on the phenomenon of 'repeat events' to redress the current trend.

Additionally, the proposed changes to the mechanisms used to feed back incident findings such that targeted lessons are channelled directly into Works Control packages, should improve the learning and hence the reduction / elimination of repeat events.

THE EXPERIENCE OF EMERGENCY SHUTDOWN OF THE VVR-c REACTOR AFTER 40 YEARS OF OPERATION

Oleg Yu. Kochnov and Yuri V. Volkov
Obninsk State Technical University for Nuclear Power Engineering

F. Lambert and Y. Volkov (eds.), Safety Improvements through Lessons Learned from Operational Experience in Nuclear Research Facilities, 47–65.

1. INTRODUCTION

The VVR-c research nuclear reactor (15 MW) has been in operation since 1964 at the Obninsk Branch of the Federal State Unitary Enterprise "NIFKhI im. L. Ya. Karpova" (L. Ya. Karpov Research Institute of Physics and Chemistry) (Fig. 1). The VVR-c is a heterogeneous, water-moderated tank reactor. The reactor was specially adapted for the performance of a broad range of studies in the field of radiation chemistry, structural and materials science research, activation analysis, neutron doping of semiconductors, etc.

Fig. 1. Overall view of VVR-c reactor.

The reactor is equipped with vertical and horizontal experimental channels of different diameters. The process complex of the reactor includes 21 hot chambers, a well-developed process transport system, a radioactive waste processing unit, etc.

2. 1. OPERATING MALFUNCTIONS AND DISRUPTIONS

There were no incidents impacting the safety of the reactor for the period from 1964 to 2004. The equipment of the complex has operated reliably in the standard mode without serious breakdowns. There were 148 unscheduled shutdowns of the VVR-c reactor during the entire operating period (see Fig. 2).

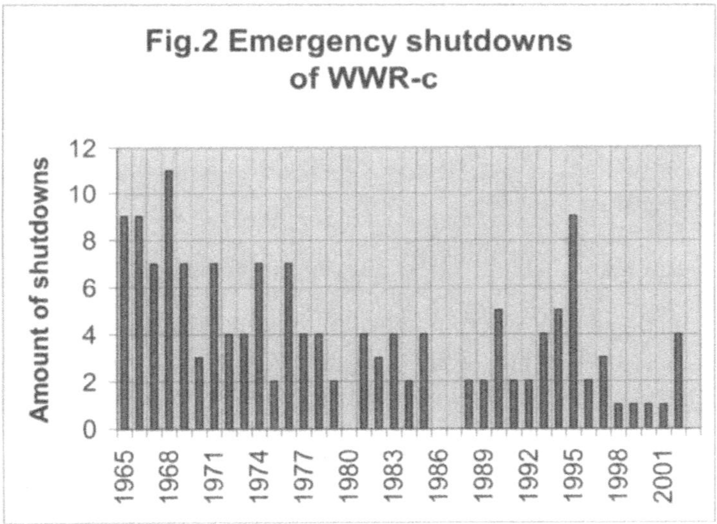

The following causes of unscheduled shutdowns of the VVR-c are worth noting:

1. voltage boosting, including transformer malfunctions – 50 (34%);
2. personnel errors:
   a) control engineer (senior reactor operator – SRO) – 13 (9%);
   b) engineer for control and protection systems, instrumentation and mechanics – 12 (8%);
3. malfunctions of experimental devices – 13 (9%);
4. malfunctions of reactor equipment:
   a) instruments – 30 (20%);

   b) unexplained cases – 24 (16%)[1];
   c) mechanical equipment – 7 (5%);
5. emergency response, with subsequent restoration of reactor power – 94 (64%);
6. emergency response, with loss of "MAK[2]" production (radioactive pharmaceutical preparation) after March 27, 1979 – 17 of 49 (35%). The average single loss of production of the "MAK" experimental channel is estimated at $5 – 15 thousand (depending upon the channel capacity).
7. emergency response, when tan emergency protection button is pressed – 15 (10%).

Having examined and analyzed emergency damping for the entire operating period of the VVR-c reactor (since 1964) [1], as well as the causes of damping, one can state:

1. The mechanical equipment of the reactor, given favorable operation and maintenance, continues to work with a minimum of malfunctions.
2. The reactor instrumentation needs to be updated.
3. The reliability of the power supply for the reactor needs to be improved.
4. Failures of experimental equipment increase the probability of accidental damping, resulting in economic losses. Reducing the number of malfunctions depends upon prompt detection of anomalies in the behavior of the reactor.
5. Training for operations personnel is one of the most important tasks in operating a reactor, since personnel are at fault in about 17% of all unscheduled reactor shutdowns[3]. The result is interruption of the experiments, loss of radioactive isotope products and, consequently, damage to the economic performance figures for the VVR-c reactor. Hence it is advisable to have a functional simulator to train personnel for each research reactor. For this purpose, a very simple simulator must be developed to simulate reactor operation in steady-state and transient operating modes to enhance the skills of personnel.

[1] The number of responses dropped to zero after 1980. Obsolete emergency protection instruments were replaced with new instruments at that time.

[2] Channel for 99Mo production.

[3] For the BOR-60 reactor (RIAR, Dimitrovgrad), 25% of unscheduled shutdowns for the entire operating period were caused by personnel errors [6].

3. PROCESSING OF EMERGENCY SHUTDOWN DATA

In a number of cases, in studying operating data or data of experiments on nuclear reactors, it is necessary to establish, for example, whether trends are present or absent in the data, the approximate form of determinate functions in the data, etc. This is especially important when there is a large spread in the operating data, and it is difficult to establish visually the presence or absence of trends, much less their form. This study proposes a simple data processing method that makes it possible to extract determinate functions from the data and avoid subjectivity in the process. Hence the following data processing algorithms were selected:

3.1 Method of Sequential Elaboration of the Model.

The essence of the method is as follows. Assume a sample $x(t_1)$, $x(t_2)$, . . . , $x(t_n)$ of volume n from the determinate function $y(t)$. This sample will include both the function $y(t)$ itself and the measurement errors $\delta(t)$, as well as, possibly, natural fluctuations $z(t)$ near $y(t)$. Hence, in general form,

$$x(t_i) = y(t_i) + \delta(t_i) + z(t_i) . \qquad (2.1.1)$$

If the theoretical function $y(t)$ is known, it can be fitted to the experimental data by a method such as the well-known least square method, and a problem can arise only in ensuring the accuracy of estimates of the parameters of the function. In a case where the theoretical function is unknown, one possible way out of the situation is to select a polynomial of the appropriate order. There are two dangers here:

1. If the order of the polynomial is too low, some important details of the function may be erroneously excluded from consideration.
2. If the order of the polynomial is too high, a situation in which details unrelated to the physical phenomenon in question, which have appeared due to measurement errors, will be taken into consideration in the function constructed is possible. Within the limit of the polynomial order $n - 1$, the polynomial will pass through all the experimental points and circumscribe all the "harrowing" experimental details.

Let

$$y_m(t) = c_0 + c_1 t + c_2 t^2 + \cdots + c_m t^m \qquad (2.1.2)$$

Theoretically, given a sufficient sample volume n, one can estimate the parameters $c_0, c_1, \dots, c_m$ for any $m < n$ based on the sample using the least square method. The problem is to select the appropriate value of m. It can be solved in various ways, but the best method appears to be (and this has been verified in practice) progressively raising m and stopping according to a definite criterion. Assume that $m = 0$. Then an average is taken,

$$c_0 = \frac{1}{n}\sum_{i=1}^{n} x(t_i)$$

Assume $m = 1$. Then a linear trend is taken. At $m = 2$, a quadratic trend is taken, etc.

Stopping is required when the deviations of $x(t)$ from the trend $y_m(t)$ are defined primarily by measurement errors. We shall consider the measurement errors to be independent and stationary (there is no error drift in the measurement method itself). Hence **the criterion for stopping the raising of the polynominal order:** it must stop when the discrepancies, i.e., $x(t_i) - y_m(t_i)$, become stationary and independent. There are criteria which make it possible to test this hypothesis [2]: the series criterion and the trend criterion, etc.

Example: Figure 3 shows damping due to faulty instruments and unexplained cases. A sequence of 36 observations of the number of cases of emergency damping is considered.

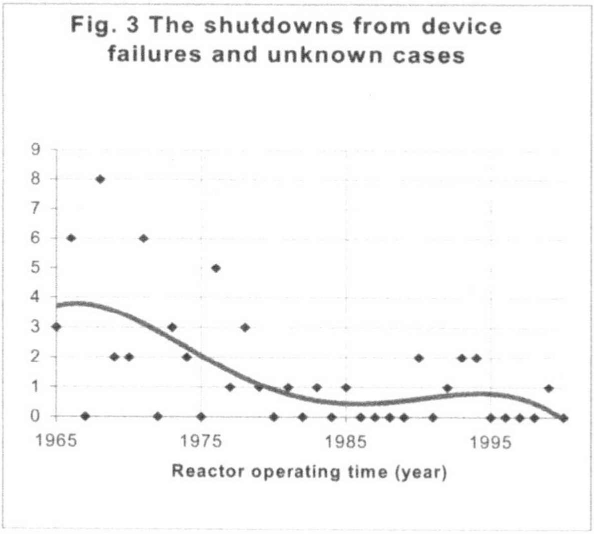

Deviations from polynomial functions of various orders are analyzed. Based on the trend criterion, we determine the number of inversions for the polynomial order from n=0. We consider the hypothesis that the observations represent independent observed values of a random variable with no trend. We consider the hypothesis at the significance level α=5%. Based on the data (see [2]), the acceptance range for the hypothesis is defined by the inequality $A_{36;1-\alpha/2}<A<A_{36;\alpha/2}$. For the case in question, 245<A<362. Table 1 presents the number of inversions as a function of the order of the approximating polynomial.

Table 1

Item	Polynomial Order	Polynomial Equation	Number of Inversions	Independence Hypothesis Accepted
1	1	y=-0.1036x+3.3889	432	No
2	2	$y=0.005x^2-0.2898x+4.5679$	427	No
3	3	$y=-0.00005x^3+0.0078x^2-0.3307x+4.7027$	436	No
4 5	4 5	y=-0.00004x4+0.0026x3-0.057x2+0.2188x+3.5614 y=0.0000001x5-0.00005x4+0.003x3-0.0626x2+0.2506x+3.5147	40 248	No Yes

The table shows that deviations from the polynomial of order n=5 satisfy the independence hypothesis for the trend criterion.

Now we shall examine data processing by the following method. The essence of the method is as follows: deviations from an approximating polynomial are plotted on a probabilistic scale, and one can see clearly that if the distribution is not normal, the polynomial order is increased, and if the deviations are distributed normally, the polynomial order is sufficient.

Example: Figure 4 shows the total number of instances of damping of the VVR-c reactor. Figure 5 gives plots of deviations on a probabilistic scale.

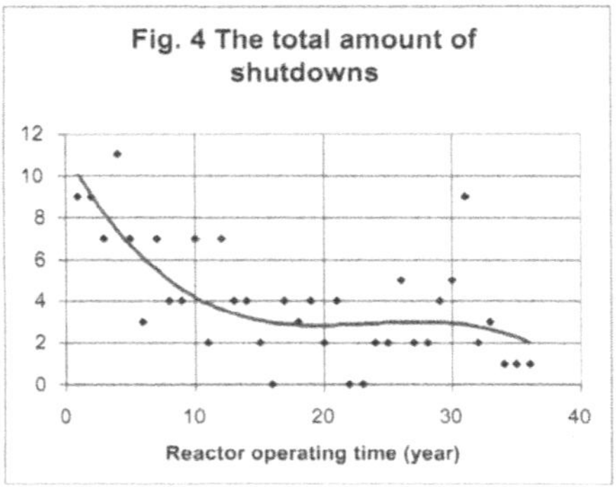

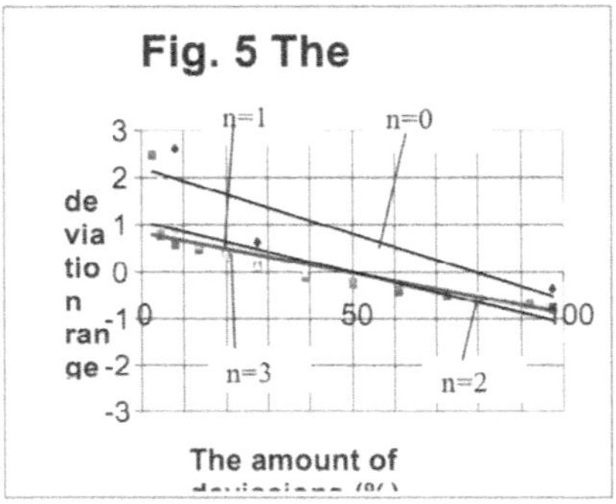

One can see that the polynomial order should be increased to n=2. Then the lines plotted on the probabilistic scale coincide, and there is no further reason to increase the polynomial order. However, this question is subjective. Having divided the intervals into smaller sections, we can attempt to raise the polynomial order further. However, it is sufficient in this case to stop at an order n=2 or n=3, which reflects the basic trend of the data row. In increasing the order, we see that the deviations also are still distributed normally, and the nature of the distribution has not changed (lines for n=2 and n=3 practically coincide).

3.2 Method of Odd Sets

Special attention is currently being devoted to problems of decision-making under complex conditions. Mathematical models have come to be widely used to describe and analyze complex economic, social and other systems. Optimization theory has created a set of methods that help to make decisions effectively with known and fixed parameters using computers. There have also been definite achievements in cases where the parameters are random variables with known distribution laws.

However, fundamental difficulties arise when the parameters of the situation are indeterminate (although not necessarily random), and when, at the same time, they strongly affect the decision results. Specialists often encounter the need for calculations when the parameters in the equations are not distinctly specified, or process information is not precise. Since determinate methods are most often used in constructing formal models, determinacy thus is introduced into situations where it is does not actually exist. Lack of precision in specifying particular calculation parameters is practically ignored, or, based on certain hypotheses and assumptions, the imprecise parameters are replaced with expert estimates or average (unweighted average) values. The irregularities that arise in this process in equations, balance ratios, etc., necessitate varying certain parameters to satisfy precisely the specified equations and obtain an acceptable result.

The theory of odd sets has been selected to define the events that have the greatest impact on emergency signals. Operating logs are kept during the operation of any facility with fissile material, and information on the functioning and malfunctions of various equipment, in particular, is recorded in the logs. It should be particularly emphasized at once that work with any database requires its own procedures, including management of database maintenance. It is well known that negligence in data input, failure to update data promptly, failure to back up critical files, failure to protect the information, etc., are the chief enemies of databases. The data recorded during the operation of the VVR-c reactor are no exception. Quite often, in the column "description and cause of malfunction," in addition to the properly recorded deviation from normal operating conditions, one finds the phrase "cause not established" or "design defect," or a description of structural failures observed in the accident ("condenser rupture," "line break," "structural displacement," etc.).

Before considering types of responses separately, we have prepared a mutual impact matrix (a special instance of the general impact matrix) for emergency signals and possible causes (Table 3). The matrix is in the form of a table, in which the rows are reactor emergency signals, and the columns are possible causes. An expert estimate is placed at the intersection of the signal and the cause. In the course of the inquiry, the expert expresses his level of confidence (a number between 0 and 1) that the specific event given is the cause of the selected emergency signal.

The numerical values presented in the mutual impact matrix correspond to the following linguistic variables given in Table 1. This table has marks on an attachment scale (analog of the Harrington desirability function [3]) for the basic (simple) linguistic variables used by the experts in the course of the inquiry (Table 2).

Table 2

Value of Linguistic Variable	Mark on Attachment Scale
Event fully impacts emergency signal	1.00 – 0.80
Event impacts emergency signal	0.80 – 0.63
Event indirectly impacts emergency signal	0.63 – 0.37
Event possibly impacts emergency signal	0.37 – 0.20
Event is not the cause of emergency signal	0.20 – 0.00

We shall analyze the mutual impact matrix and consider the relationships between events and emergency signals in order of priority.

Table 3 Mutual Impact Matrix

Causes of automatic protection function / Emergency signals	Personnel error	Instruments	Unexplained cases	Voltage boosts	Experimental devices	Reactor equipment	Time of year	Natural phenomena (thunderstorms)	Building operation in liquid radioactive waste processing	Time of day 8:00 to 20:00	Time of day 20:00 to 8:00
Water level in reactor (DSR)[4]	0.3	0.8	0.3	0.1	0.1	0.2	0.0	0.0	0.0	0.0	0.0
Water level in reactor (MED)[5]	0.3	0.8	0.3	0.1	0.1	0.2	0.0	0.0	0.0	0.0	0.0
Capacity 120% Period < 10 sec Electronic equipment malfunction[6]	0.4	1.0	0.9	1.0	0.0	0.1	0.0	0.5	0.0	0.0	0.0
Water flow rate Primary circuit	0.2	0.5	0.3	0.5	0.0	0.4	0	0.0	0.0	0.0	0
Wate pressure primary circuit	0.2	0.5	0.3	0.5	0.0	0.4	0	0.0	0.0	0.0	0
Water flow rate Secondary circuit	0.7	0.5	0.3	0.5	0.0	0.4	0.5	0.0	0.2	0.2	0
Water pressure secondary circuit	0.7	0.5	0.3	0.5	0.0	0.4	0.3	0.0	0.2	0.2	0
Water temperature at reactor inlet	0.8	0.5	0.3	0.2		0.2	0.8	0.0	0.0	0.0	0
ΔT between reactor water inlet and outlet	0.2	0.5	0.3	0.2	0.0	0.0	0.0	0.0	0.0	0.0	0
Main circulating pumps of primary circuit off	0.0	0.0	0.3	0.8	0.0	0.4	0.0	0.2	0.0	0.0	0
100%	1.0	0.0	0.3	0.0	0.2	0.1	0.0	0.0	0.0	0.0	0.8

[4] The VVR-c is a tank reactor; the core is in a cylindrical tank. The tank is filled with desalinized water. The water level H≈5.2 m. It is measured with a DSR level gauge (range 0 – 7 m).

[5] For an alternative measurement of the water level, an MED level gauge is installed in the reactor tank (range 0 – 2 kgf/cm^2). The instrument serves for the second independent emergency protection channel [30].

[6] Combined signal of 3 components with 1 of 1 majority logic.

immersion of control rod											
No 110V voltage on control and protection system busbars	0.0	0.0	0.3	0.0	0.0	0.0	0.0	0.0	0.0	0.0	0
Water flow rate through experimental channel (EK 4-9)[7]	0.0	0.5	0.3	0.8	0.9	0.0	0.0	0.0	0.0	0.0	0

[7] Experimental channel EK4-1 serves for irradiating targets with 235U for producing the 99Mo isotope. The signal "drop in water flow rate through channel" is connected to the reactor emergency protection.

To determine the events with the greatest impact on emergency signals, an attachment function is plotted:

$$\mu_k(x) = \frac{\sum_{i=1}^{n}\left(\sum_{j=1}^{m} x_{jk} \Big/ m\right)}{n} \tag{2.1.3}$$

where j = 1 . . . m is the number of experts;
k is the cause of functioning of automatic protection;
i = 1 . . . n is the number of emergency signals;
x_{jk} is the assessment of the j expert of the k cause of functioning of automatic protection.

The number of emergency signals n=13. In this case, the decision concerning the appearance of an emergency signal under the effect of a particular event (k=11) is made by two experts (m=2); naturally, to improve the "quality" of the assessment, we need a larger number of qualified experts.

From calculations performed according to formula (2.1.3), we obtain the value of the attachment function:

$\mu_{\text{personnel error}}(x) = 0.37;$
$\mu_{\text{instruments}}(x) = 0.47;$
$\mu_{\text{unexplained cases}}(x) = 0.35;$
$\mu_{\text{voltage boosts}}(x) = 0.40;$
$\mu_{\text{experimental devices}}(x) = 0.10;$
$\mu_{\text{reactor equipment}}(x) = 0.22;$
$\mu_{\text{time of year}}(x) = 0.12;$
$\mu_{\text{natural phenomena}}(x) = 0.05;$
$\mu_{\text{building operation for processing liquid radioactive waste}}(x) = 0.03;$
$\mu_{\text{time of day from 8:00 to 20:00}}(x) = 0.03;$
$\mu_{\text{time of day from 20:00 to 8:00}}(x) = 0.06.$

One can see from the calculation results that four types of events have the greatest numbers of effects on emergency signals: personnel errors, voltage boosts, instruments and unexplained cases; the latter events are qualified as "event possibly impacts emergency signal." Since the number of instances of emergency damping in unexplained cases has declined to zero as a result of updating of equipment, only three types of events remain as "event indirectly impacts emergency signal." One of these, "personnel errors," has prompted the design of the "Operator Information Support Complex" (KIPO) system [4].

4. OBJECTIVES OF KIPO DEVELOPMENT FOR VVR-TS

The objectives of the development of the KIPO system and the range of specialists who might make use of KIPO information are listed below:

1. training of personnel, especially SROs, before qualifying examinations (such as in approval of new personnel for work);
2. training of operations personnel, such as reactor control engineers and reactor shift supervisors;
3. playing out reactor operation scenarios (for training personnel from the process engineer group, for example);
4. most accurate determination of the "acceptable idle time" and "iodine pit depth" in emergency reactor shutdown for orientation of the shift supervisor and process engineer group;
5. issuing of "advice and recommendations" for the SRO during reactor operation.

Even partial accomplishment of these objectives will facilitate VVR-c operation considerably and reduce the probability of errors in the work of personnel, which will produce a significant economic effect for the RIPC Branch. The operating reliability of the research reactor will also improve, and the "safety culture" [5] among personnel will be augmented.

5. GENERAL DESCRIPTION OF THE KIPO

5.1 KIPO Structure

The overall structure of the KIPO for the VVR-c reactor is presented below. The interactive part of the KIPO consists of the following units (see Fig. 6):

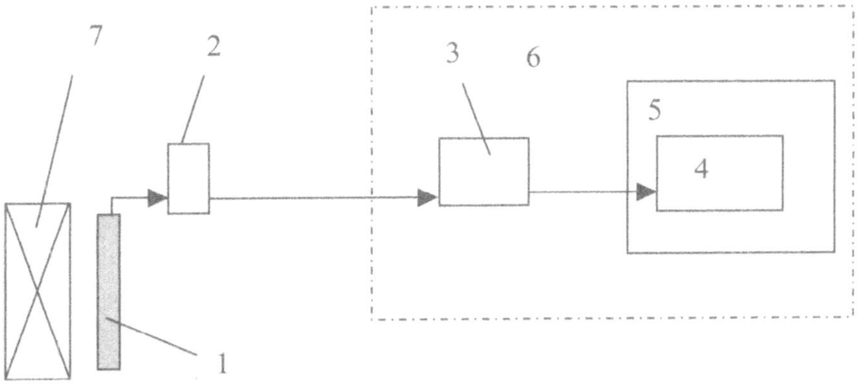

Fig. 6. Block diagram of the interactive part of the KIPO.

1. KNK-15 ionization chamber
2. Preamplifier of current signal from chamber
3. ZPT2-01 instrument complex
4. analog-to-digital converter
5. personal computer
6. main control room
7. VVR-ts reactor core

The current signal passes from the stationary KNK-15 ionization chamber in the vertical channel in direct proximity to the VVR-ts reactor core (standard reactor instrumentation complex) to the preamplifier in the central hall (CH). From the preamplifier, the signal goes to the ZPT2-01 unit[8] located in the reactor main control room (MCR). From the ZPT2-01 output, the signal is fed to the A/D converter of a personal computer in the MCR. The signal is converted to digital code by the A/D converter and used

[8] Period current protection.

in the program responsible for determining the reactor status. In simulator mode, the current signal can be simulated by a current source. The current source is a rectifier which can smoothly regulate the output load. Variation in the reactor power, as well as emergency shutdown, can be simulated by variation in the current level.

5.2 KIPO Program Operating Algorithm

The program for determining the reactor status is made up of several blocks. The structure and the connections between blocks within the program are implemented as follows (see Fig. 7).

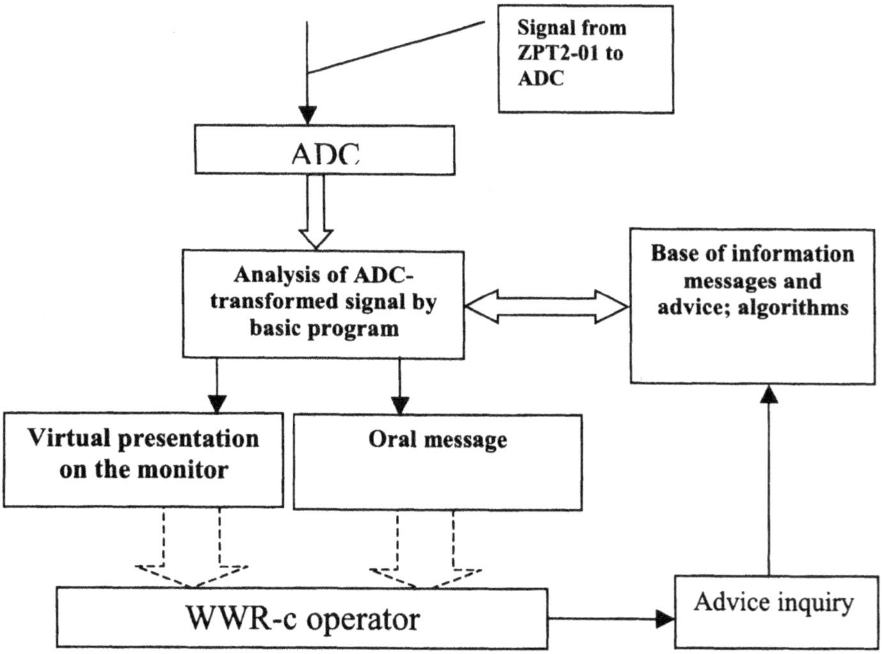

Fig. 7. Connections between blocks in the KIPO program.

The signal from the A/D converter is used by the main program for calculating the current reactor status. Models of point kinetics of the processes that go on in the reactor core, as well as extremely simple models of the primary and secondary circuits and the cooling tower, are built into the program. The A/D converter is queried every second to adjust the data on the neutron flux in the reactor, which is required for stable operation of the program. Processing of readings with a frequency an order lower is generally adequate, since the refresh rate for screen details is once every minute. To obtain more reliable data, however, the data are averaged.

$$n_{cp} = \frac{\sum_{i=1}^{60} n_i}{60}$$; where n_i is the chamber current level.

The program results are displayed graphically on a monitor screen in simple and convenient form for the reactor operator (Fig. 8).

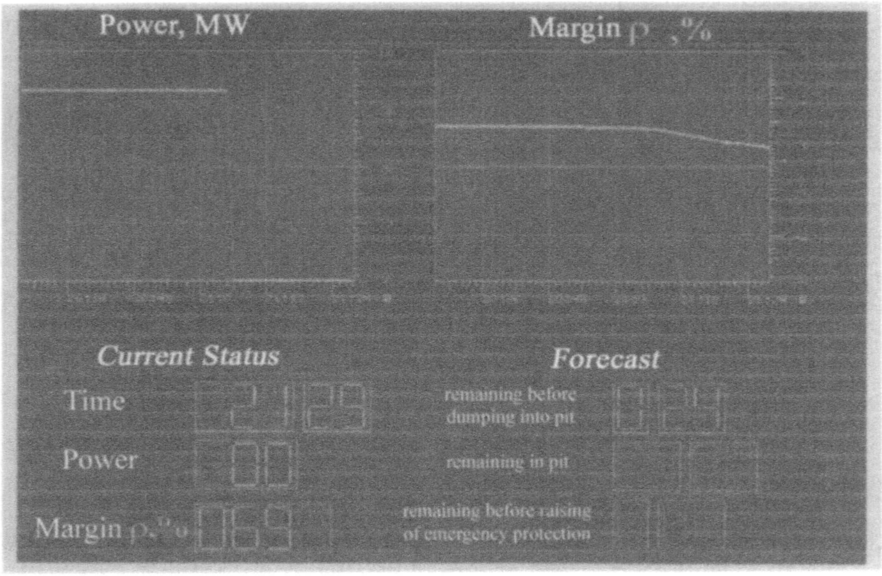

Fig. 8. KIPO operation detail.

Voice messages are also generated for more comprehensive perception of the information. Programs responsible for advice and informational messages run parallel to the main program. When the operator wants to hear advice or a message, the appropriate mode is activated via the interface. A series of informational messages are always generated regardless of the user's preference. An informational message on the computer monitor is duplicated by a voice message through the computer speaker.

5.3 Results of Work on the KIPO Design

The KIPO system is in operational condition. The user interface is being modified constantly. The system is used often – at least once a week – in simulator mode. After long-term (six months') test operation in the basic operating mode, it will be possible to address the issue of its permanent use in the main reactor operating mode.

5.4 Deployment on the VVR-c Reactor

Tests of the first version were satisfactory. KIPO has acquitted itself well in simulator mode. It has proved especially useful to the process group in planning reactor operating conditions during the summer. KIPO has received a positive response from young SROs and young specialists taking examinations.

Another advantage of KIPO over full-scale simulators is the fact that it does not require the special panels and control boards used by standard full-scale simulators, since it is made up of a personal computer and a minimal set of input signals.

5.5 References

1. Kochov, O. Yu., and Yu. V. Volkov. "Analysis of Data on Responses of VVR-ts Reactor Automatic Protection." Obninsk: *Yadernaya energetika*, No. 2, 2002.
2. Bendat, J., and A. Pearsall. "Measurement and Analysis of Random Processes." Moscow: Mir, 1971.

3. Kandel, A., and U. J. Bayatt. "Odd Sets, Odd Algebra and Odd Statistics." Proceedings of the American Society of Electronics Engineers, v. 66, 1978, No. 12, pp. 37-61.
4. Koziyev, I. N., O. Yu. Kochnov, E. S. Stariznyy and Yu. V. Volkov. "VVR-c Operator Information Support Complex. Experience of Development of the First Version." Obninsk: *Yadernaya energetika*, No. 2, 2000.
5. OPB IR-2000 [General Regulation for Ensuring Research Reactor Safety]. Moscow: Energoatomizdat, 2000.
6. Collected Papers of the International Scientific and Technical Conference "Research Reactors: Science and High Technologies." Dimitrovgrad, 2002, v. 2, part 1.

MINOR INCIDENTS DURING THE DECOMMISSIONING OF PROTOTYPE OPERATION AND RESEARCH FACILITIES OF THE KARLSRUHE RESEARCH CENTER

Wolfgang Pfeifer and Horst Goenrich
Forschungszentrum Karlsruhe GmbH, Geschäftsbereich Stilllegung Nuklearer Anlagen, Hermann-von-Helmholtz-Platz 1, 76344 Karlsruhe, Germany

Abstract: It is a declared objective of both the federal government and the competent state authorities that safety of nuclear facilities in the Federal Republic of Germany is to be given highest priority and that no compromises shall be made with respect to their safe operation. The operators of nuclear facilities are obliged to guarantee safe operation of their facilities by themselves and in their own responsibility in principle. Operators of nuclear facilities in Germany therefore undertake to analyze the causes of reportable events with a view to further increase the safety and reliability of their plants.

The nuclear facilities of the Forschungszentrum Karlsruhe are in various stages of dismantling. Apart from dismantling, however, also safe remaining operation of the plants must be ensured. This results in very high requirements on the safety culture and in particular on man's acting in interaction with the organization.

1. INTRODUCTION

On behalf of the Federal Ministry of Education and Research (BMBF), the Forschungszentrum Karlsruhe carried out research and development work (R&D) for advanced reactors and facilities of the nuclear fuel cycle. For this purpose, four larger prototype reactors and a reprocessing plant were built and operated. It was demonstrated that nuclear facilities can be operated in an economically efficient manner with all safety-relevant

F. Lambert and Y. Volkov (eds.), Safety Improvements through Lessons Learned from Operational Experience in Nuclear Research Facilities, 67–78.

boundary conditions being observed. Particular attention was paid to the parallel development of radiation protection measures and an optimum organization structure to ensure the safety of the operations personnel and environment. Very early, it was agreed on criteria for reporting special incidents with the competent licensing authorities. These criteria served as a basis of the presently valid Nuclear Safety Officer and Reporting Ordinance (AtSMV) for facilities according to Art. 7 of the Atomic Energy Act (AtG).

The four **prototype reactors** and reprocessing plant owned by the Forschungszentrum Karlsruhe have already been dismantled completely or are in an advanced state of dismantling[1].

The **Karlsruhe Reprocessing Plant (WAK)** is being dismantled in three phases[2]:

- Dismantling of the **main process building**; all installations have been removed. Decontamination of the cell walls has started.
- Construction and operation of the **vitrification facility (VEK)** for the solidification of about 70 m^3 high-active liquid waste concentrate (HAWC); this facility will be taken into operation in 2005.
- Dismantling of the **HAWC storage facilities** upon the completion of vitrification.

At the WAK in particular, organization has to meet high requirements, as is not only required to dismantle contaminated facilities, but also to safely handle the HAWC solutions and to operate the vitrification facility. In this connection, handling and communication of minor incidents are of crucial importance.

2. NUCLEAR SAFETY OFFICER AND REPORTING ORDINANCE

Since 1975, reportable events in nuclear facilities in the Federal Republic of Germany have been reported to the nuclear supervisory authorities in accordance with the valid federal reporting criteria and compiled in a central list. Since October 1992, the obligation of the operators to report such events to the supervisory authorities has become legally binding with the adoption of the **Nuclear Safety Officer and Reporting Ordinance (AtSMV)**[3].

It is the purpose of the administrative reporting procedure to supervise and improve the safety status of the nuclear facilities in Germany.

The incidents reporting office of the **Federal Radiation Protection Authority (BfS)**[4,5] lists and documents all incidents in nuclear facilities that have been reported to the BfS among others. These incidents are then evaluated on behalf of the **Federal Ministry of the Environment, Nature Protection, and Reactor Safety (BMU).**

According to the AtSMV, reportable events are subdivided into the following categories:

Category S Immediate report; reporting period – immediately
Category E Urgent report; reporting period - within 24 hours
Category N Normal report; reporting period - within 5 days
Category V Only for facilities under construction.

Irrespective of the administrative reporting procedure according to the AtSMV, reportable events are additionally classified by the operators of the nuclear facilities using the International Atomic Energy Agency's (IAEA) **"International Nuclear Event Scale" INES.** As a rule, reportable events of the AtSMV category "**N**" correspond to the INES level "**0**".

Furthermore, the **Gesellschaft für Anlagen- und Reaktorsicherheit (GRS, organization for the safety of facilities and reactors)**, also on behalf of the BMU, evaluates all reports of incidents in nuclear facilities that have become known worldwide via the "Incident Reporting System (IRS)" of the OECD and IAEA[6]. If the incident is of relevance to German facilities, GRS issues a **forwarding message**. This forwarding message is sent to the state supervisory authorities, the experts called in by them, and the respect-tive operators. The operators check this forwarding message for potential transferability and applicability to their facilities. The result of this check and the state of implementation of the GRS recommendations are reported by the operators to the competent state supervisory authorities. In Baden-Wuerttemberg, this is the Ministry of the Environment and Transport (UVM). Upon evaluation of the operators' reports and statements, the supervisory authority decides on whether further checks are required and the countermeasures proposed and taken are sufficient.

3. INTERACTION OF MAN, TECHNOLOGY, AND ORGANIZATION

Safety of a nuclear facility is determined by the **factors of man, technology, and organization** (MTO system). However, these factors are not independent of, but influence each other.

Technology is influenced by man's acting. At the interface of man and technology, for instance, operation of the facility, inspections to be made, and maintenance measures have to be considered.

The organization with its written in-plant procedures shall ensure that the plant is operated safely and inspections are made systematically and comprehensively. Such procedures also help prevent human errors. The operations and work flows prescribed by internal regulations (operations manual, inspections manual, etc.) represent major organizational factors that influence technology. Human performance is affected by organizational factors, such as written instructions, work permits, and work plans. They have to be designed such that human performance and reliability are supported and errors are avoided, if possible. Human capabilities and actions in nuclear facilities are of particular relevance to safety. There will always be unforeseeable situations, to which technology alone cannot react in an adequate manner. In such cases, human problem solution competence and flexibility are required, with all internal regulations being observed.

Moreover, unwritten social rules that are reflected by attitudes and the conduct of superiors, assistants, and colleagues are of particular importance. Together, they form the **safety culture of the operator**.

The International Nuclear Safety Advisory Group (INSAG) defines safety culture as follows: "Safety culture is the entirety of features and attitudes of organizations and individuals, as a result of which highest priority is given to safety-related issues of nuclear power plants getting that attention that corresponds to their relevance".

A high safety culture is characterized by a questioning basic attitude, a well reflected and careful proceeding, and an intensive and open communication.

The **organization** is understood to comprise the persons responsible on the individual hierarchy levels (management), their stipulations and regulations as well as the internal procedures. The organizational structure of a nuclear facility is outlined in the operations manual (BHB). Here, also the tasks and competencies of the management are defined. For optimization, it is important how the organization succeeds in learning from the experience gained. This organizational learning should not be blocked by blaming.

The term **human factor (HF)** denotes all potential influencing factors at the man-machine interface, which may occur when operating the facilities. Systematic listing, analysis, and assessment of HF-relevant events and their consideration by the plant organization are referred to as HF system. The HF system depends on voluntary reports of the employees who rely on no disadvantages resulting for them when they make such reports.

4. ORGANIZATION OF THE DECOMMISSIONING DIVISION

Responsibility for dismantling of the nuclear facilities of MZFR, KNK, and WAK of the Forschungszentrum Karlsruhe lies with the **Decommissioning Division** (Fig. 1).

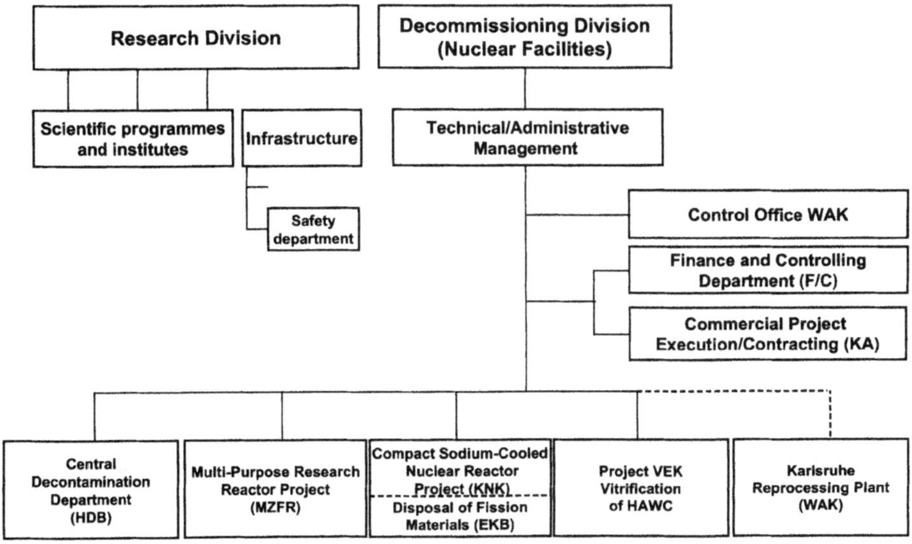

Figure 1. Organization chart of the Forschungszentrum Karlsruhe

For the nuclear facilities of KNK, MZFR, and HDB operated by the **Decommissioning Division**, the control tasks according to the Atomic Energy Act (AtG) and the Radiation Protection Ordinance (StrlSchV) are assumed by the **Central Safety Department (HS)** of the **Research Division**.

As the Central Safety Department is directly responsible to the Chairman of the Executive Board, the **four-eyes principle** in matters related to safety is maintained irrespective of the operator's interests.

As the owner of the reprocessing plant, the Forschungszentrum Karlsruhe has ordered the **WAK operation company** to decommission and dismantle the plant. The technical director of the WAK and the Chairman of the Executive Board of the Forschungszentrum Karlsruhe are responsible for matters related to radiation protection. As the Forschungszentrum Karlsruhe also is one of the holders of the license of the WAK, the Ministry of the Environment of Baden-Wuerttemberg considered it necessary to establish a **control office (auditing)** that assumes control tasks concerning plant safety, radiation protection, and waste disposal in the WAK. It is one of the tasks of this control office **to check the handling of reportable events in the WAK.** In this connection, particular attention is paid to the identification of causes, a complete presentation of the course of events without any contradictions, the determination of noticeable weaknesses (technical, administrative, organizational, personal), and the lessons to be learned from the reportable event.

The **plant organization** used for monitoring the facilities for the storage of the high-active liquid waste concentrate (HAWC) at WAK has not changed in principle after the stop of reprocessing operation. The present organizational structure of the WAK comprises four departments and 12 sub-departments that are responsible to the **plant management** (plant manager):
Operation (plant management, shift supervisor)
Monitoring (analysis, radiation protection, residue and waste disposal)
Technology and maintenance (mechanics, electrical engineering, media)
Dismantling (several partial project heads)

With the completion of vitrification operation only can the present plant organization be transferred to a more project-oriented dismantling organization (as in case of the reactors) with a smaller number of interfaces. Then, it will also be possible to adapt the organization to the changing tasks while dismantling the nuclear facility.

For the reactors KNK and MZFR, a **project-oriented organization** was established (Fig. 2). Project management, planning, and procurement are separated from the responsibility for the operation and dismantling of the plant. This organizational structure allows for a clear delimitation between dismantling and organization-management, project- and safety-management.

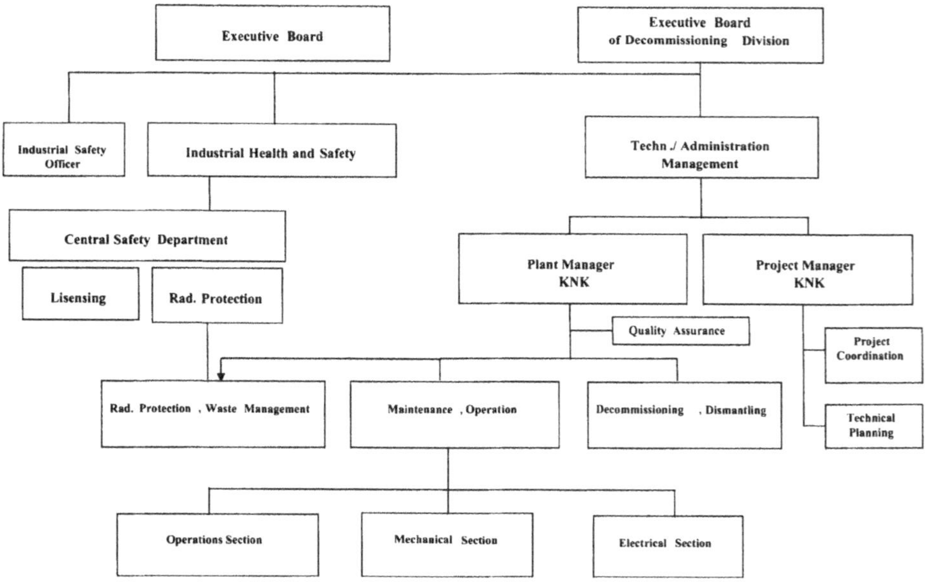

Figure 2. Organization chart of the KNK Decommissioning Project.

5. ANALYSIS OF REPORTABLE EVENTS

The causes of reportable events in principle lie in the areas of man, technology, and organization. As the quality of technical systems has improved constantly, technical defects of safety and operation systems have decreased. At the same time, human errors and deficiencies of the internal organization have gained importance.

Reports about all reportable events of a year are issued[4,5] by the Federal Radiation Protection Authority (BfS) for facilities for the fission of nuclear fuels and facilities for the supply and disposal of nuclear fuels.

The results obtained from evaluating the causes of all events in German nuclear facilities covered by the BfS reports of the years 1999 to 2003 and the corresponding evaluation results for the facilities of the Forschungszentrum Karlsruhe are summarized in Tab. 1.

		Nuclear Power Plants		*Nuclear Fuel Supply and Disposal*	
Category		**Germany**	**MZFR and KNK**	**Germany**	**WAK**
AtSMV	**S**	**2 *** (0.3 %)	**0**	**0**	**0**
	E	**20** (3.1 %)	**0**	**1** (0.5 %)	**0**
	N	**624** (96.6 %)	**13**	**198** (99.0 %)	**146** (99.3 %)
	V	**0**	**0**	**1** (0.5 %)	**1** (0.7 %)
INES	**2**	**2 *** (0.3 %)	**0**	**1** (0.5 %)	**1 **** (0.7 %)
	1	**25** (3.9 %)	**0**	**3** (1.5 %)	**3 ***** (2.0 %)
	0	**619** (95.8 %)	**13**	**196** (98.0 %)	**143** (97.3 %)
Causes *					
Component defects		**296** (45.8 %)	**11** (84.6 %)	**100** (50.0 %)	**90** (61.2 %)
Mode of operation		**34** (5.3 %)	**0**	*not determined*	*not determined*
Design, planning		**75** (11.6 %)	**0**	**13** (6.5 %)	**1** (0.7 %)
Production, assembly		**66** (10.2 %)	**0**	*not determined*	*not determined*
Operation, maintenance		**136** (21.1 %)	**2** (15.4 %)	*not determined*	*not determined*
Human errors		*not determined*	*not determined*	**42** (21.0 %)	**37** (25.2 %)
Insufficient regulations		*not determined*	*not determined*	**4** (2.0 %)	**4** (2.7 %)
Cause not identified		**0**	**0**	**11** (5.5 %)	**3** (2.0 %)
Others		**7** (1.1 %)	**0**	**15** (7.5 %)	**12** (8.2 %)
Under investigation		**32** (5.0 %)	**0**	**15** (7.5 %)	**0** (0 %)

Table 1. Evaluation of the BfS Annual Reports of 1999 to 2003 for Facilities in Germany

These evaluations allow the following conclusions to be drawn:

According to the BfS evaluation, about 21% of the reportable events in German nuclear power plants are caused by **operation and maintenance errors** and, hence, by human errors or deficiencies of the organization. At the facilities of the Forschungszentrum Karlsruhe that are presently being dismantled, about 15% of the reportable events are caused by human errors.

Regarding facilities for the supply and disposal of nuclear fuel, 21% of the reportable events may be attributed to human errors. At the Karlsruhe Reprocessing Plant WAK, the respective share amounts to about 25%.

It is most probable, however, that **human errors and deficiencies** of the organization also play a far greater role for other causes of reportable events, e.g. mode of operation, planning, assembly, insufficient regulations, than is generally assumed. Frequently, several factors lead to reportable events, such that a distinction of individual cause categories is much too general. Hence, the assessment of reportable events by the BfS may only give a qualitative overview.

The findings from detailed analyses of reportable events, reports of special incidents, and audit results of the control office have to be communicated to all employees of the Decommissioning Division in the sense of a learning organization. Only this will allow for a further improvement of the safety of dismantling and remaining operation of the nuclear facilities of the Forschungszentrum Karlsruhe.

6. EXAMPLES OF REPORTABLE EVENTS THAT OCCURRED IN THE DECOMMISSIONING DIVISION OF THE FORSCHUNGSZENTRUM KARLSRUHE

Four reportable events that occurred in the Decommissioning Division of the Forschungszentrum Karlsruhe shall now be analyzed for their causes. The measures taken and lessons learned from the reportable events shall be presented. All reportable events were ranked category N according to AtSMV and 0 according to INES.

6.1 Drop of a Load in the Controlled Area

What happened:

On June 25, 2003, a contaminated steel plate of about 2500 kg in weight dropped down, while it was transported by a forklift to a container in the controlled area.

Evaluation:

According to the expert, this event is a result of deficiencies in safety culture with following reasons:

- Responsible persons of various organizational units have decided in favor of an unapproved transport path. The transport path taken was permitted for the transport of contaminated parts in drums only, but not for steel plates packed into foil.
- The forklift used was not qualified for this task. Again, the decision to use a forklift was approved of by various organizational units.

- Deficiencies of the quality assurance system have become obvious: The method selected for the identification and marking of loads and approved of by the responsible persons obviously is highly susceptible to human errors (mixing up of labels).

Measures taken:

- The plant manager issued a work instruction, in which the determination of the weight and the marking of the residues to be transported are defined more clearly.
- This changed procedure is communicated to the dismantling, radiation protection, and residue logistics staff in additional training sessions.

6.2 Deficiencies in the Execution of Work on Sampling Devises for High-active Fission Product Solutions (HAWC)

What happened:

When determining the air flow rate of air blowers of various sampling devices in the facility for the "storage of high-active fission product solutions", the alarm of the local dose rate monitor in a shielded area was set off during measurement. The monitor in the accessible area did not react.

The alarm of the dose rate monitor was caused by HAWC liquid having entered the system due to the vacuum in the suction line, which had not been taken into consideration. For work on the HAWC sampling system, no work permit had been issued.

The alarm of the dose rate monitor and the corresponding signal in that area were not reported to the shift supervisor and superior officers due to an incorrect assessment of the cause and its significance.

Evaluation:

In view of the numerous violations of in-plant procedures and inspection regulations on several organizational levels, the expert sees deficiencies in safety culture. The behavior of the staff was not characterized by a questioning attitude and safety-oriented decision-making.

The staff is trained concerning the requirements on dismantling activities in the main process building. Operational switching processes during the monitoring of HAWC storage operation are rare. Operational routine is lacking in this area.

Measures taken:

- Improvement of the specific technical knowledge especially in the HAWC storage area by specialized training.
- Work on facilities related to HAWC that requires a work permit must be approved of separately by the plant manager after technical settlement.

- Changing the responsibility of the sampling devices from the analytical group to the maintenance department.

6.3 Wound Contamination of an External Employee in the Research Center's Facility for the Processing of Radioactive Waste (HDB)

What happened:

In the scrapping plant of the Central Decontamination Department (HDB), a drum compact got stuck in the press discharge unit. A worker removed the material by hand, slipped off, and the upper side of his ring finger was cut slightly. The wound was contaminated.

Evaluation:

This accident is due to the following causes:

- The tools available for loosening the stuck compact were not used.
- The regulation, according to which stuck compacts must not be removed manually from the press, was violated.

Measures taken:

- Procurement of protective gloves which provide for a better protection even in case of erroneous actions of the staff
- General safety instruction in the light of this event and additional, case-related safety instructions

This accident also revealed additional deficiencies in the first aid by the staff to treat open wounds in the controlled area. For this reason, instruction of the staff as to how to congest a wound properly was initiated and continuously check of the inventory of the first-aid cabinet with respect to hoses and clamps for wound congestion (together with the Central Safety Department and the Medical Department of the Research Center).

7. OUTLOOK AND RECOMMENDATION

Forschungszentrum Karlsruhe has already gathered vast experience from the complete dismantling of nuclear facilities in all phases as well as from complex waste treatment activities during operation and dismantling of these facilities. No serious reportable events occurred that had repercussions both inside and outside of the plants.

By constant adaptation of organizational processes and a consistent and transparent work preparation and execution, we have succeeded in reliably avoiding larger incidents and a violation of the limit values given in the Radiation Protection Ordinance.

Based on the experience gathered, the following recommendations can be made:

- **Operations Manual**

  The analysis of events leads to modifications of the operations manuals. In the course of time, however, these manuals that govern all processes become too complex and can no longer be handled. Updating of the operations manual and its adaptation to the progress of dismantling with a removal of obsolete procedures are of particular importance. Over-regulation must be avoided.

- **Organization**

  Compared to plants in operation, dismantling projects are carried out with a super proportionally high deployment of external staff. The organization should provide for simple and clear responsibilities. Management of the work contracts is of particular importance. Contractors have to take measures to increase and improve safety.

- **Maintenance of Competence**

  For the maintenance of competence, the plant management is obliged to inform the responsible members of the staff recurrently about the current state of the plant and the pertinent procedures. In this respect, the safety specifications laid down in the operations manual and its maintenance, radiation protection, and reporting sections and compliance with the operational requirements are of particular importance. Furthermore, the operations instructions for daily routine activities have to be taken into consideration.

  Maintenance of competence shall include specific trainings on facilities to be commissioned.

- **Constant Improvement**

  A forum should be established, where **free exchange of opinion** may take place **with the aim of constant improvement**. For this, it is required to establish a work climate that is open for such a feedback and free of blaming. A good example is the Industrial Safety Committee established at the Forschungszentrum Karlsruhe.

  For a constant improvement process, procedures and rules have to be set up. Under the in-house suggestion scheme, each employee may submit proposals for improvement. By the payment of premiums, incentives are given to the employees to take part.

- **Audit and Quality Assurance**

  By a control office, random controls (audits) of compliance with the provisions of the Atomic Energy Act and licenses for ensuring a safe remaining operation and dismantling of the plant are made. Among others, compliance with operational regulations for emission monitoring, recurrent inspections, the accounting of radioactive substances, and the declaration of radioactive residues is checked. Implementation of the recommendations made by this control office results in improvements of the internal work flows for dismantling. Unclear interfaces between technical departments and sub-departments can be specified more precisely or eliminated. Hence, work of this control office contributes to improving safety by the implementation of its recommendations.

8. REFERENCES

1. W. Pfeifer, Decommissioning of Major Nuclear Facilities of Forschungszentrum Karlsruhe, EAES – European Atomic Energy Society, 27./28.5.2002, Sintra/Portugal
2. W. Pfeifer, J. Fleisch, Forschungszentrum Karlsruhe; G. Katzenmeier, Wiederaufarbeitungsanlage Karlsruhe; Decommissioning of the Karlsruhe Reprocessing Plant (WAK), Atomic Energy Agency Technical Specialists Meeting, IAEA Headquarters, Vienna/Austria, 2–4 April 2003
3. Verordnung über den kerntechnischen Sicherheitsbeauftragten und über die Meldung von Störfällen und sonstigen Ereignissen (Atomrechtliche Sicherheitsbeauftragten- und Meldeordnung – AtSMV); zuletzt geändert am 18. Juni 2002; BGBl. IS. 1869, 1906
4. Jahresberichte des Bundesamtes für Strahlenschutz über Meldepflichtige Ereignisse in Anlagen zur Spaltung von Kernbrennstoffen in der Bundesrepublik Deutschland, Jahrgänge 1999 bis 2003
5. Jahresberichte des Bundesamtes für Strahlenschutz über Meldepflichtige Ereignisse in Anlagen zur Kernbrennstoffver- und Entsorgung in der Bundesrepublik Deutschland, Jahrgänge 1999 bis 2003
6. Konzeption für die staatliche Aufsicht über die baden-württembergischen Kernkraftwerke; UVM, März 2001

ANALYSIS OF INFORMATION ON INCIDENTS AT RESEARCH NUCLEAR PLANTS IN RUSSIA

V. V. Kalygin, V. N. Fedulin and P. S. Smirnov

Russian Federation State Scientific Center Research Institute of Atomic Reactors – SSC RF RIAR

Abstract: The paper contains the results of analysis of disruptions at research nuclear plants (RNPs) in Russia in 2002. Data are given on the breakdown of disruptions by enterprises operating RNPs and by categories defined by the Regulation on Procedures for Investigation and Accounting of Disruptions in the Operation of RNPs. The direct and root causes of disruptions which are encountered most often and corrective actions intended to reduce the number of disruptions at RNPs are covered. It is demonstrated that a downward trend can be seen in the number of disruptions in RNP operation.

F. Lambert and Y. Volkov (eds.), Safety Improvements through Lessons Learned from Operational Experience in Nuclear Research Facilities, 79–86.

1. INTRODUCTION

Research nuclear reactors are a basic source of intense neutron fluxes. Research reactors are of major importance for the development of many basic sciences and nuclear engineering. All the countries in which national concepts have been established for the development of nuclear power have research nuclear plants (RNPs) with various capacity levels and for various purposes. More than 250 research reactors are currently active in 58 countries, including about 30 in Russia. The experience of the operation of research reactors amounts to 12,000 reactor-years, which is significantly greater than the corresponding figure for nuclear power plants – 10,000 reactor-years.

Analysis of the experience of operating research reactors helps in identifying the causes of disruptions and extracting useful lessons. This information is used by the organizations that operate and design reactor plants to improve RNP safety.

An industry Center for Collection and Analysis of Information on RNP Safety (CAI RNP) was established at the SSC RF RIAR in 1998. The CAI RNP has developed an information system (IS) for collection and processing of data on disruptions in the operation of RNPs. The IS covers 22 organizations of 8 ministries and departments of Russia which operate about 100 RNPs (including critical and subcritical test beds) and submit information to the CAI RNP on disruptions in RNP operation.

The International Atomic Energy Agency (IAEA) implemented a reporting system for incidents at research reactors (IRSRR) in 2000 under the rubric of the RNP safety program. The IRSRR was created for prompt delivery of technical information and information on lessons extracted from incidents at research reactors to the nuclear community. The incident cause analysis method is used in the IRSRR. The experience accumulated by the IAEA during work with the reporting system for disruptions at NPPs (NEA) was used in developing the IRSRR. The IRSRR functions as an international forum for sharing RNP operating experience.

2. SITUATION IN RF

There were 41 disruptions recorded at RNPs of Russia in 2002. The overall breakdown of disruptions in RNP operation by enterprises of Russia is shown in Fig. 1.

It is worth mentioning that there were no disruptions at mothballed research reactors, reactors undergoing reconstruction, reactors being decommissioned or reactors under construction in 2002. One disruption was recorded at a critical test bed.

The distribution of the specific disruption index for enterprises of Russia (the ratio of the number of disruptions to the number of active research reactors of the enterprise) is shown in Fig. 2.

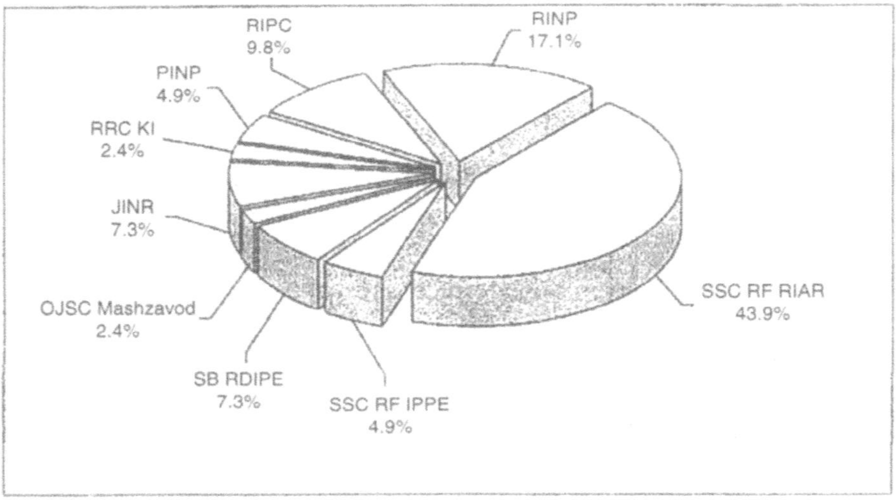

Fig. 1. Overall breakdown of disruptions in RNP operation by enterprises in Russia.

Note: RIPC – Research Institute of Physics and Chemistry; RINP – Research Institute of Nuclear Physics;
PINP – Petersburg Institute of Nuclear Physics; RRC KI – Russian Research Center, Kurchatov Institute;
JINR – Joint Institute for Nuclear Research; OJSC Mashzavod – Mashzavod Open Joint-Stock Company;
SB RDIPE – Sverdlovsk Branch of the Research and Development Institute of Power Engineering; SSC RF

IPPE – Russian Federation State Scientific Center – Institute of Physics and Power Engineering; SSC RF

RIAR – Russian Federation State Scientific Center Research Institute of Atomic Reactors.

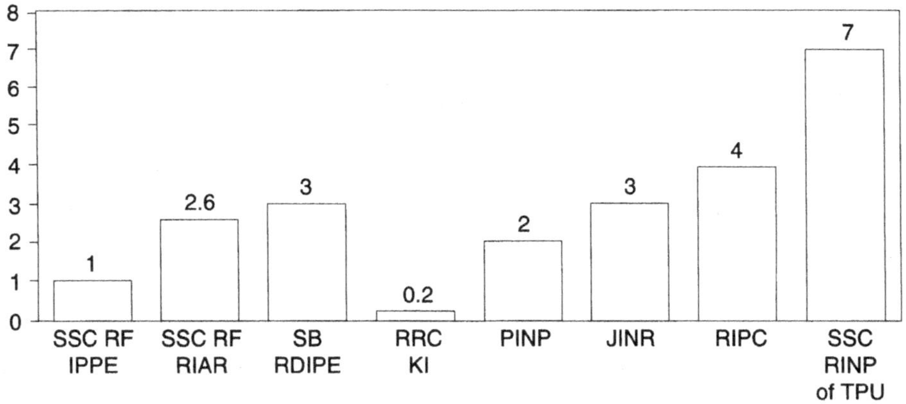

Fig. 2. Specific disruption index for RNP operation.

Note: SSC RINP of TPU – State Scientific Center Research Institute of Nuclear Physics of Tomsk Polytechnic University.

An analysis of disruptions in RNP operation was performed by the CAI RNP based on "Investigation Reports on Disruptions in the Operation of RNPs," prepared in accordance with the Regulation on Procedures for Investigation and Accounting of Disruptions in the Operation of RNPs, document NP-027-01.

The breakdown of disruption in RNP operation by categories is shown in Fig. 3.

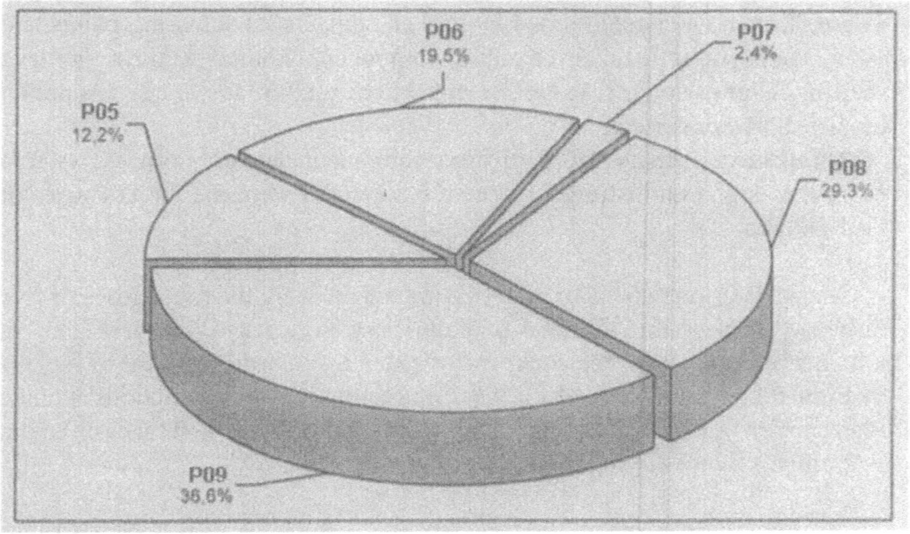

Fig. 3. Breakdown of disruptions in RNP operation by categories.

3. ANALYSIS OF EVENTS

Disruptions in RNP operation are distributed as follows, according to the categories provided for in standard document NP-027-01:

- 15 disruptions (36.6%) involving RNP shutdowns caused by anomalies in the operation of external electric power systems (category P09);
- 12 disruptions (29.3%) involving RNP shutdowns caused by anomalies in the operation of monitoring and control systems (category P08);
- 8 disruptions (19.5%) involving RNP shutdowns caused by personnel errors (category P06);
- 5 disruptions (12.2%) involving RNP shutdowns caused by operating failures of process and (or) electrical equipment and (or) piping (category P05);

- 1 disruption (2.4%) involving RNP shutdown caused by operating failure of an experimental device (category P07).

The most common direct causes of component failures were malfunctions of instrument systems (8 failures) and electrical malfunctions (5 failures). These failures were conditioned by broken contacts (at soldering points and other connections) and irreparable component failures (relays, voltage stabilizers, transistors, resistors, microswitches, etc.), as well as insulation breaks and other defects.

Mechanical damages to a process equipment actuator, process system components and an instrument system component were the direct causes of four failures.

Manufacturing defects of system components were the root causes of five failures. The same number of failures were caused by problems in administrative control, management or planning of work (failure to replace instruments promptly based on wear and age). Corrective actions include replacement of obsolete and worn out monitoring and measurement instruments by modifying the RNP designs.

Shortcomings in repair and maintenance procedures were the root cause of failures ranking next in significance (4 failures). Corrective actions include making the necessary changes to repair and maintenance procedures for components and increasing the frequency of maintenance in the light of wear (age).

The most frequent direct causes of errors of workers (personnel) were errors in operational switching (5 failures), deviations from operating documentation (4 failures) and poor monitoring of the performance of operational switching by shift supervisors (3 errors). Errors in operational switching, as a rule, were made in transient reactor operating modes, which attests to the need for improving the quality of training for personnel.

Functional and analytical simulators (FASs) for VK-50 and BOR-60 reactors are currently being developed at the training center (TC) of the SSC RF RIAR. The development and use of FASs to simulate various transient modes of operation of the reactors makes possible substantial improvement in the quality of practical training for operations personnel and the overall operating safety of research reactors.

4. ANALYSIS OF CAUSES

Deviations from operating documentation were conditioned mainly by the inadequate training of personnel on safety culture issues. Safety culture calls for both strict fulfillment of all the requirements of operating documentation (including organizational issues) and precise execution of a superior's operational commands by subordinate personnel, with special attention to the performance of difficult and critical operations (including supervision of their performance). Random action on a protective element, arbitrary switching and improper actions in maintenance were also conditioned by inadequate training of personnel on safety culture issues.

Corrective actions in regard to workers (personnel) mainly involved renewing the theoretical knowledge of the personnel who made errors by unscheduled training and unscheduled examinations. Corrective actions in regard to other areas were purely technical or organizational in nature.

Of the 15 disruptions involving anomalies in the operation of external electrical systems:

- 10 disruptions were conditioned by anomalies in the operation of external electrical systems located outside the operating organizations;
- 5 disruptions were conditioned by anomalies in the operation of systems located on the grounds of the operating organizations.

The causes of anomalies in the operation of external power supply systems of the RNPs were:

- damage to the systems caused by personnel errors during operational switching (3 disruptions) and during construction work (4 disruptions) and by shortcomings in the organization of the operation of the systems – construction of an entertainment center in protection zone of a high-voltage power line was allowed (1 disruption);
- short circuiting in a 6 kV high-voltage cable caused by a hidden manufacturing defect (1 disruption);
- operating malfunction of high-voltage equipment (1 disruption).

The causes of anomalies in the operation of external electrical systems were not identified in 5 cases of disruptions (all involved systems located outside the operating organizations).

The consequences of anomalies in the operation of external power systems for the RNPs in all cases were shutdowns of the RNPs by automatic functioning of the automatic protection systems without violating operating limits or conditions for safe operation.

5. CONSEQUENCES

Analysis of the disruptions in the operation of RNPs in Russia indicated the following:

- 41 disruptions were recorded at RNPs in Russia in 2002;
- 40 disruptions occurred at active research reactors, and one occurred at a critical test bed. There were no disruptions at research reactors undergoing reconstruction, mothballed reactors, reactors being decommissioned or reactors under construction, or at subcritical test beds;
- all the disruptions were classified as categories P05 – P09. There were no disruptions classified as categories A01, A02 or P01 – P04;
- all the disruptions in the operation of RNPs in Russia were classified as level 0 (nonessential for safety) on the INES scale.

Table 1 presents comparative data on the causes of disruptions in the operation of RNPs in Russia for 2000, 2001 and 2002.

Table 1
Breakdown of Disruptions in RNP Operation in Russia

Item	Cause of Disruption	Period		
		2000	2001	2002
1.	RNP system (component) failures	29	32	18
2.	Anomalies in the operation of external power systems	24	12	15
3.	Errors of RNP workers (personnel)	9	10	8
	TOTAL	62	54	41

From the table, one can see a declining trend in the number of disruptions in RNP operation.

PRACTICES IN GOVERNMENT REGULATION OF THE SAFETY OF RESEARCH NUCLEAR PLANTS IN RUSSIA

Practices of the French Atomic Energy Commission

Aleksandr Sapozhnikov
Federal Environmental, Technological and Nuclear Inspection Service (Rostekhnadzor)

1. INTRODUCTION

Since the collapse of the Soviet Union, there have been significant changes in the structure of government executive agencies with the authority to regulate the safety of nuclear and radiation hazard facilities in the Russian Federation (Russia).

F. Lambert and Y. Volkov (eds.), Safety Improvements through Lessons Learned from Operational Experience in Nuclear Research Facilities, 87–110.

Practices in government regulation of the safety of civilian research nuclear plants[1] in a developing market economy are the subject of this paper; the paper includes a brief survey of the following issues:

* the government system for safety regulation in the use of atomic energy;
* principles for government supervision of nuclear and radiation safety;
* the system for collection, analysis and processing of operating information to improve the safety of research nuclear plants.

2. THE GOVERNMENT SYSTEM FOR SAFETY REGULATION IN THE USE OF ATOMIC ENERGY

The table shows the changes in government agencies for regulating the safety of the use of atomic energy in the former USSR and then in Russia. The system for government regulation of nuclear and radiation safety in Russia has developed in stages. A brief description of these stages is given below.

In the initial phase of the development of the nuclear industry, all research nuclear plants, as well as other facilities for the use of atomic energy, were under the control of a specially created ministry of the USSR. The ministry made centralized decisions on issues both of control of the use of atomic energy and of safety regulation in the use of atomic energy. There was on-site monitoring of nuclear and radiation safety. Funding for measures to ensure and improve the safety of active research nuclear plants, as a rule, was provided on a carry-over basis. The imbalance in the atomic energy usage policy for the initial period was one of the main causes of accidents at research nuclear plants and led to the problem of radioactive waste disposal. The system of regulations governing the safety of research nuclear plants in place at the time did not conform in regard to scope or

[1] The term "research nuclear plant" will be used hereinafter to refer to facilities and complexes with research nuclear reactors (RRs) and critical and subcritical nuclear test beds intended to use neutrons and ionizing radiation for research purposes.

content to international practices calling for the application of legislative principles to achieve safety.

The USSR State Committee for Supervision of Working Safety in Nuclear Power Engineering (GAEN), formed in 1983, was the country's first government regulatory agency in the field of the use of atomic energy; GAEN began developing the IAEA ideology based on the systematic development of standards and regulations for nuclear and radiation safety. Technical substantiation of safety (TOB) documents, including a review of design and hypothetical accident scenarios, issues in establishing the life expectancy of equipment and performing refurbishment, measures to bring research nuclear plants into compliance with new safety requirements, and issues of the decommissioning of research nuclear plants, were developed for each research reactor. The main objective in developing TOB documents was to document safety analyses to confirm that the reactor in question could operate according to established safety procedures, standards and regulations, and that its use could be carried out safely.

A whole range of executive agencies in addition to GAEN were established in the USSR with special targets in regard to governing safety in the use of atomic energy, such as construction, sanitary and hygienic aspects of radiation safety, technical safety, fire safety, environmental protection and physical protection, among others. Certain coordination of the activities of these executive agencies was required.

Further centralization of functions of government regulation of safety in branches of the industry was effected in 1990. The USSR State Committee for Supervision of Industrial and Nuclear Safety (Gospromatomnadzor SSSR) was formed based on the USSR nuclear regulatory authority Gosatomenergonadzor and the USSR mining and industrial inspection authority Gosgortekhnadzor to designate new licensing (permit) procedures for phases and types of work to be performed in industry and nuclear power engineering.

After the collapse of the USSR, pursuant to Decree No. 249 of the President of the RSFSR dated December 3, 1991, and on the basis of Presidential Order No. 137-rp dated December 31, 1991, all facilities for the use of atomic energy in the Russian Federation, regardless of their nature or departmental affiliation, were transferred to the purview of the newly formed State Committee for Supervision of Nuclear and Radiation Safety of the

President of the RSFSR (Gosatomnadzor RSFSR, which was later renamed Gosatomnadzor of Russia). Gosatomnadzor RSFSR was assigned the functions of managing and conducting government regulation and supervision of the safe use of atomic energy, nuclear materials and radioactive substances for peaceful and defense purposes in the Russian Federation. The Regulation on the Russian Federal Supervision of Nuclear and Radiation Safety (Gosatomnadzor of Russia) was approved by Order No. 283-rp of the President of the Russian Federation dated June 5, 1992, with amendments approved by Orders Nos. 636-rp and 350-rp of the President of the Russian Federation dated September 16, 1993, and July 26, 1995. The Regulation defined the objectives, jurisdiction and powers of Gosatomnadzor of Russia and gave a list of types of activities that could be conducted by subjects of entrepreneurial activity and enterprises, regardless of forms of ownership, only on the basis of a permit (license) from agencies of Gosatomnadzor of Russia. Government supervision of nuclear and radiation safety in the development, manufacture, testing, use, storage and disposal of nuclear weapons and nuclear power plants for military purposes was later assigned to the Ministry of Defense of Russia by Order No. 350-rp of the President of the Russian Federation dated July 26, 1995.

From mid-1993 to July 14, 1997, Rosatomnadzor of Russia issued provisional permits[2] for types of activities involving the use of atomic energy. Governing documents establishing the procedures and conditions for obtaining provisional permits from Gosatomnadzor of Russia until a federal law on the use of atomic energy could be adopted and enacted were developed and implemented in 1994. The procedure of issuing provisional permits was terminated as of July 14, 1997, and licensing of types of

[2] A provisional permit is a document issued by Gosatomnadzor of Russia certifying the competence of the enterprise and conferring the right to conduct a specific type of operations or to render services in the field of the use of atomic energy on the condition of ensuring the safety of the facilities for the use of atomic energy and the work performed.

activities involving the use of atomic energy has been conducted since January 1, 1998.

A three-tiered system of federal executive agencies is currently taking shape in Russia, including: federal ministries, federal services and federal agencies. The Federal Environmental, Industrial and Nuclear Regulatory Authority (Rostekhnadzor) has been called upon to combine the operations involved in resolving issues of the regulation of safety for a variety of industrial facilities in a single executive body. The federal executive agencies authorized to conduct government regulation of safety in the use of atomic energy in Russia are:

- Federal Environmental, Industrial and Nuclear Regulatory Authority (Rostekhnadzor) – in regard to environmental, technological, industrial, fire, nuclear and radiation (technical aspects) safety;
- Federal Service for Protection of Consumers Rights and Human Welfare (Rospotrebnadzor), under the jurisdiction of the Ministry of Health and Social Development (Minzdravsotsrazvitiya of Russia) – in regard to radiation safety (sanitary and hygienic aspects);
- Russian Federation Ministry of Civil Defense, Emergencies and Management of Natural Disasters (MChS of Russia) – in regard to fire safety.

The regulation on the Federal Environmental, Industrial and Nuclear Regulatory Authority was approved by Russian Federation Government Resolution No. 401 dated July 30, 2004. The operations of Rostekhnadzor and MChS of Russia are supervised by the Government of the Russian Federation. Figure 1 shows the structure of Rostekhnadzor.

Rostekhnadzor currently performs government regulation of safety at 80 research nuclear plants of 21 operator organizations (research nuclear plant proprietors) of various ministries and agencies. Under federal law, research nuclear plants in Russia are federal property.

The main tasks of the state system for regulating the safety of research nuclear plants in Russia are given in Fig. 2 under three regulatory components: standards, licensing and inspection. The safety regulatory pyramid and the Russian authorities responsible for developing the respective regulatory documents are shown in Fig. 3. A complete list of legal regulations and documents used for government regulation of the safety of research nuclear plants is presented in the List of Basic Regulatory Instruments and Regulatory Documents Used by Gosatomnadzor of Russia for Government Regulation of Safety in the Field of the Use of Atomic Energy (P-01-01-2003), *Vestnik Gosatomnadzora Rossii* [*Journal of Gosatomnadzor of Russia*], No. 2, 2003.

Activities in the field of the use of atomic energy for peaceful and defense purposes (with the exception of activities involving nuclear weapons and nuclear power plants for military purposes) are governed by Federal Law No. 170-FZ "On the Use of Atomic Energy" enacted November 21, 1995.

The legal grounds for ensuring radiation safety are defined in Federal Law No. 3-FZ "On Radiation Safety of the Public" enacted January 9, 1996.

Safety assurance requirements for research nuclear plants are contained in federal rules and regulations (FNP). Federal rules and regulations are developed under the rubric of the federal target program "Nuclear and Radiation Safety of Russia for 2000-2006" and within the framework of international cooperation. The FNP system has the following structure:

1. General provisions.
2. Site selection, design, construction.

2.1. Safety requirements for the facility as a whole.

2.2. Requirements for safety-critical systems.

2.3. Requirements for equipment (components).

3. Operation and decommissioning.

   3.1. Requirements for personnel.

   3.2. Safety requirements for work and services.

   3.3. Decommissioning.

   3.4. Emergency preparedness.

4. Requirements for substantiation of safety.

General provisions include the following documents:

- NRB-99 Radiation Safety Standards;
- NP-033-01 Basic Safety Regulations for Research Nuclear Plants, *Journal of Gosatomnadzor of Russia*, No. 2, 2002;
- NP-042-02 Requirements for the Quality Assurance Program for Research Nuclear Plants, *Journal of Gosatomnadzor of Russia*, No. 1, 2003.

According to NP-033-01:

A research nuclear plant satisfies safety requirements if its radiation impact on workers (personnel), the general public and the environment in normal operation and disruptions in normal operation, including design accidents, does not result in exceeding established exposure doses for workers (personnel) or the general public, or the standards for emissions (discharges) or environmental concentrations of radioactive substances, and is limited in beyond-design accidents. The adequacy of the physical barriers and technical and organizational measures for deep multi-echelon protection at research nuclear plants must be substantiated in the design and presented in the substantiation of safety report (OOB) for the research nuclear plant. Assessment of the compliance of research nuclear plants with the nuclear and radiation safety requirements for such plants is performed in the following form:

- licensing of types of activities;
- government supervision of the process of ensuring compliance with regulatory requirements.

Assessment of the compliance of products intended for use at facilities for the use of atomic energy is performed in the following form:

- acceptance;
- mandatory certification (for mass-produced products).

Licensing of the activities at research nuclear plants is conducted in accordance with the Federal Law On the Use of Atomic Energy on the basis of the Regulation on the Licensing of Activities in the Field of the Use of Atomic Energy, enacted by Russian Federation Government Resolution No. 865 dated July 14, 1997, and regulatory documents of Rostekhnadzor (Gosatomnadzor of Russia).

To obtain a license, the operator (research nuclear plant proprietor) must submit documents to substantiate its readiness to conduct the declared type of activity to Rostekhnadzor in compliance with the established requirements. An expert review of the documents for substantiation of safety may be initiated on specific subject matter. The need for an expert review is determined by Rostekhnadzor in the process of examining the documents for substantiation of safety. The expert review is conducted by an expert organization duly licensed by Rostekhnadzor (Gosatomnadzor of Russia) under a contract with the operator organization. Rostekhnadzor conducts an inspection on problematic safety issues and to confirm the reliability of the information submitted by the applicant. A license is issued for a period of at least 3 years and is accompanied by conditions for validity of the license. The fulfillment of these conditions is monitored by territorial agencies of Rostekhnadzor. If necessary, the conditions for validity of the license may be revised according to established procedures.

In the initial stage of the licensing of research nuclear plants with a long operating life, at the request of the operating organization, licenses were issued for periods less than 3 years for bringing the research nuclear plants into compliance with current requirements.

The conditions for validity of the license have the following structure:

- area in which the license is valid;
- general requirements;
- requirements for information and reporting on activities;
- requirements for information on irregularities;
- requirements for nuclear materials and radioactive substances control and accounting and for providing physical protection for research nuclear plants, nuclear materials and radioactive substances;
- special requirements.

In accordance with the Federal Law On the Use of Atomic Energy, Gosatomnadzor of Russia established the procedures for issuing work permits to the management and personnel of research nuclear plants. The list of positions of personnel at facilities for the use of atomic energy requiring permits from Gosatomnadzor of Russia for work in the field of the use of atomic energy was confirmed by Russian Federation Government Resolution No. 240 dated March 3, 1997. The conditions for validity of an issued permit, which define the boundaries of the safety responsibilities of the specific functionary at the research nuclear plant and in the operator organization, are an integral part of that permit. All research nuclear plant supervisors currently have permits, and Rostekhnadzor is continuing to refine the required procedures.

3. PRINCIPLES FOR GOVERNMENT SUPERVISION OF NUCLEAR AND RADIATION SAFETY

The activities conducted by the federal executive agency for monitoring and supervision in the field of the use of atomic energy include:

- acquisition and analysis of information on the status of nuclear and radiation safety;
- organization and performance of inspections;
- making the necessary decisions and imposing compulsory measures and sanctions when violations of the requirements for ensuring nuclear and radiation safety are discovered.

The management of government supervision of the safety of research nuclear plants is based on the following principles:

- independence of Rostekhnadzor (Gosatomnadzor of Russia) of other government agencies and organizations whose activities involve the use of atomic energy in the performance of is functions;
- delimiting of the responsibility of the parties that conduct operations in the field of the use of atomic energy and government supervision of the safety of the use of atomic energy;
- openness of government supervision – ensuring that information on the status of nuclear and radiation safety and on government supervision of safety is accessible, unless the information includes state or commercial secrets;
- interaction with other agencies involved in government regulation of safety in the use of atomic energy;

- reasonable actions in conducting government supervision – oversight activities should not unreasonably limit the activities of operator and other organizations for ensuring safety;
- a varied approach to forms of supervision, depending upon potential hazards, the level of safety assurance achieved by the research nuclear plant and the condition of its structures, systems and components.

The supervision principles set forth above are embodied in the governing documents and inspection procedures of Gosatomnadzor of Russia and preserve continuity in the oversight practices of Rostekhnadzor.

To assess the activities of the license holder, inspections are performed at the research nuclear plant and within the operator organization as a whole. The inspections may be complex, targeted or operational according to the scope of the issues to be investigated. The frequency of the inspections: for complex inspections of operator organizations managed by the central staff – at least once every 5 years. The frequency of targeted and operational inspections depends on the capacity of the research nuclear plant. For research nuclear plants with a capacity of 1 MW or higher, the frequency of targeted inspections of the research nuclear plants, conducted, as a rule, by inspectors on site at territorial agencies, is at least once a year, and the frequency of operational inspections is according to the inspector's work schedule, but at least once a quarter. Depending upon the potential hazard level of the research nuclear plant and the deficiencies in safety assurance for the plant, the number of inspections performed at one research nuclear plant in a year will vary from 3 to 15.

The types of administrative penalties and the standards for imposing compulsory measures and sanctions for violations are defined in federal laws: No. 68-FZ "On Administrative Responsibility of Organizations for Violation of Laws in the Field of the Use of Atomic Energy," enacted May 12, 2000, and No. 195-FZ "On Enactment of the Russian Federation code of Administrative Offenses," enacted December 30, 2001.

A transition from normative regulation of the safety of facilities for the use of atomic energy in accordance with the requirements of federal rules and regulations to safety regulation based on the requirements of technical procedures is going on at present. Scheduled development of federal rules and regulations are to be completed before the end of 2005, and the development of technical procedures is to be organized in accordance with Federal Law No. 184-FZ On Technical Regulation enacted December 27, 2002. The law defines the conditions for creating a consistent approach to safety assurance for various industrial facilities and quality assurance for their products.

The implementation of technical procedures in the field of the use of atomic energy will involve integration of the forms and principles of government supervision in the areas of environmental, technical, fire, nuclear and radiation safety of facilities for the use of atomic energy.

4. SYSTEM FOR COLLECTION, ANALYSIS AND PROCESSING OF OPERATING INFORMATION TO IMPROVE THE SAFETY OF RESEARCH NUCLEAR PLANTS

In accordance with the requirements of NP-033-01, operator organizations shall:

- develop and implement measures to prevent repeat violations of limits and safe operating conditions with the same causes;
- arrange for the collection, processing, analysis, systematization and storage of information on irregularities in the operation of research nuclear plants for the entire operating life of the plants, as well as prompt delivery of such information to other organizations according to established procedures.

The results of analysis of operating information are the basis for informing the public concerning the status of nuclear and radiation safety of research nuclear plants and implementing operating management measures to improve the safety of the plants:

- refining the design of the research nuclear plant and bringing the plant into compliance with current requirements;
- making addenda to substantiation of safety reports;
- defining the list of accident initiating events;
- developing an operating quality assurance program for the research nuclear plant;
- sharing experience in the operation of research nuclear plants and training of personnel.

The structure of the national system for reporting irregularities in the operation of research nuclear plants and analyzing other operating information is shown in Fig. 4. The system has three reporting levels: 1) operator organization level; 2) regulatory agency level; 3) international level. The nature of the information is different at each level, according to the different functions and tasks of the information users. The operator organizations submit to Rostekhnadzor reporting documents which are the basis for assessing the status of nuclear and radiation safety of the research nuclear plants. Rostekhnadzor turns over to the IAEA system for reporting on irregularities in the operation of research reactors (IAEA IRSRR) reports on the irregularities in the operation of research nuclear plants in Russia which are most significant with respect to safety.

Reporting requirements, including operating information, are based on the requirements of the following documents:

- Regulation on Investigation and Accounting Procedures for Irregularities in the Operation of Research Nuclear Plants (NP-027-01), *Vestnik, Journal of Gosatomnadzor of Russia*, No. 2, 2001;
- contents of the annual report of the operator organization on assessment of the status of nuclear and radiation safety of the research nuclear plants (RB-025-03).

Safety problems of research nuclear plants can be classified as organizational, technical or regulatory problems based on the results of analysis of the operating information.

The main unresolved organizational issues are as follows:

- The Government of the Russian Federation has not specified for a number of operator organizations the agencies for control of the use of atomic energy which have full responsibility for the safety of research nuclear plants under the Federal Law On the Use of Atomic Energy.

- The experimental research nuclear power plant facilities of scientific centers have almost no commissions to solve safety problems of nuclear engineering and the nuclear industry.
- Attrition of specialized enterprises and manufacturers of components for research nuclear plants is going on.
- Aging and attrition of qualified personnel are going on, and the flow of young specialists to research nuclear plants is declining due to the drop in the prestige of the profession and low pay.

The following are causes of the main technical problems of the safety of research nuclear plants in Russia:

- aging of the equipment and research nuclear plants, and insufficient study of methods for estimating and substantiating the life expectancy of equipment (components) and structural elements of research nuclear plants;
- lack of technology for processing certain types of fuel elements and fuel assemblies of research nuclear plants, and the fact that the elements and assemblies are kept at temporary storage facilities for long periods of time;
- lack of utilization technologies for unique research nuclear plants in the rehabilitation of areas of scientific centers at which an unhealthy radiation situation had developed in the early stages of activities involving the use of research nuclear plants;
- a tendency to accumulate spent (irradiated) nuclear fuel and radioactive wastes (RAW) at temporary storage facilities located in the areas of scientific centers due to sharply rising expenses for transportation and processing at specialized enterprises;
- issues in refurbishing obsolete plants for processing radioactive wastes and building new RAW disposal sites at scientific centers with their own RAW processing plants and disposal sites.

The key objectives in safety regulation for research nuclear plants are:

- generating a balanced strategy for executive agencies that manage the use of atomic energy and executive agencies that regulate the safety of research nuclear plants with respect to national policy on the use of experimental research nuclear plant facilities, and strengthening the mechanism of government regulation of the expenses of scientific centers for acquiring fresh fuel for research nuclear plants, hauling away RAW and spent fuel assemblies, and decommissioning research nuclear plants;
- implementing IAEA recommendations incorporated in the "Code of Behavior for the Safety of Research Reactors";
- developing technical procedures for nuclear and radiation safety of facilities for the use of atomic energy in accordance with the Federal Law On Technical regulation;
- developing the licensing procedure for activities of education centers for specialist training and skills enhancement for personnel of research nuclear plants and the management of operator organizations;
- developing an interagency information system for assessment of irregularities in the operation of research nuclear plants and emergency response.

5. LESSONS OF PRACTICAL WORK IN GOVERNMENT REGULATION OF THE SAFETY OF RESEARCH NUCLEAR PLANTS IN RUSSIA

Most of the research nuclear plants in Russia are located in cities or settlements. They were designed and commissioned in the 50's through the 70's, when there was no body of regulations in place on the design and operation of nuclear and radiation hazard facilities. Issues of the long-term storage of spent nuclear fuel, decommissioning of research nuclear plants, and nuclear materials and radioactive substances accounting and physical protection had not been studied in depth.

Work to develop a body of laws and regulations on safety in the use of atomic energy has been going on since the organization of the first agency for government regulation of nuclear and radiation safety in Russia. The current national system for safety regulation in the use of atomic energy is made up of three components: normative, licensing and supervision. The system of laws and regulations in the Russian Federation includes a practically complete body of requirements for the safety of research nuclear plants in the light of IAEA approaches to regulating nuclear and radiation safety in the field of the use of atomic energy. The body of safety requirements is adequate for the potential hazard level of research nuclear plants and is oriented toward research reactors with steady-state neutron flux, pulsed research reactors, critical nuclear test beds, subcritical nuclear test beds and other subcritical nuclear plants.

The proper level of safety culture at facilities for the use of atomic energy can be achieved only on the basis of balanced, coordinated development of national legislation for all types and directions of activities in the field of the use of atomic energy, including new technologies and the disposal of radioactive wastes and spent nuclear fuel.

Significant resources are required to achieve a reasonably high level of nuclear and radiation safety of research nuclear plants. In the Russian Federation at the present time, however, the market has no commercial interest in solving specific problems of nuclear and radiation safety of research nuclear plants, and the scientific centers lack the ability to include all the expenses required to maintain safety in the unit cost of scientific developments or commercial products (generation of heat, electric power, radioactive isotopes). Accordingly, strengthening of the mechanism for government regulation of the expenses of scientific centers for acquiring fresh fuel for research nuclear plants, hauling away RAW and spent fuel assemblies, decommissioning research nuclear plants and training specialists for research nuclear plants is urgently needed to improve the safety of the use of research nuclear plants.

The regulatory agency should organize and provide for the functioning of systems for monitoring research nuclear plants when emergencies arise (emergency response).

Table. Agencies for Government Regulation of Safety in the Use of Atomic Energy in the Former USSR and the Russian Federation

USSR	
1963-1970	On-site monitoring by the Central Inspectorate of the Boiler Inspection Service of the USSR Ministry of Medium Machine Building and the 3rd Main Directorate of the USSR Ministry of Health
Resolution of the USSR Council of Ministers No. 879-302 dated October 22, 1970	The USSR State Industrial and Mining Safety Inspection Committee (Gosgortekhnadzor SSSR) was charged with government regulation of safety issues in the operation of equipment at nuclear power plants and experimental and research nuclear reactors
Decree of the Presidium of the Supreme Soviet of the USSR of July 19, 1983	The USSR All-Union State Committee for Supervision of Nuclear and Radiation Safety (Gosatomnadzor SSSR) was formed
June 27, 1989	Reorganization of Gosgortekhnadzor SSSR and Gosatomenergonadzor SSSR into the USSR State Committee for Supervision of Industrial and Nuclear Safety (Gospromatomnadzor SSSR)
Russian Federation	
Decree of the President of the RSFSR No. 249 dated December 3, 1991	The State Committee for Supervision of Nuclear and Radiation Safety of the President of the RSFSR (Gosatomnadzor of Russia) was formed
Decree of the President of the Russian Federation No. 314 dated March 9, 2004	Renamed the Federal Nuclear Inspection Service
Decree of the President of the Russian Federation No. 649 dated May 20, 2004	On reorganization of the structure of executive agencies and organization of the Federal Environmental, Technological and Nuclear Inspection Service (Rostekhnadzor)

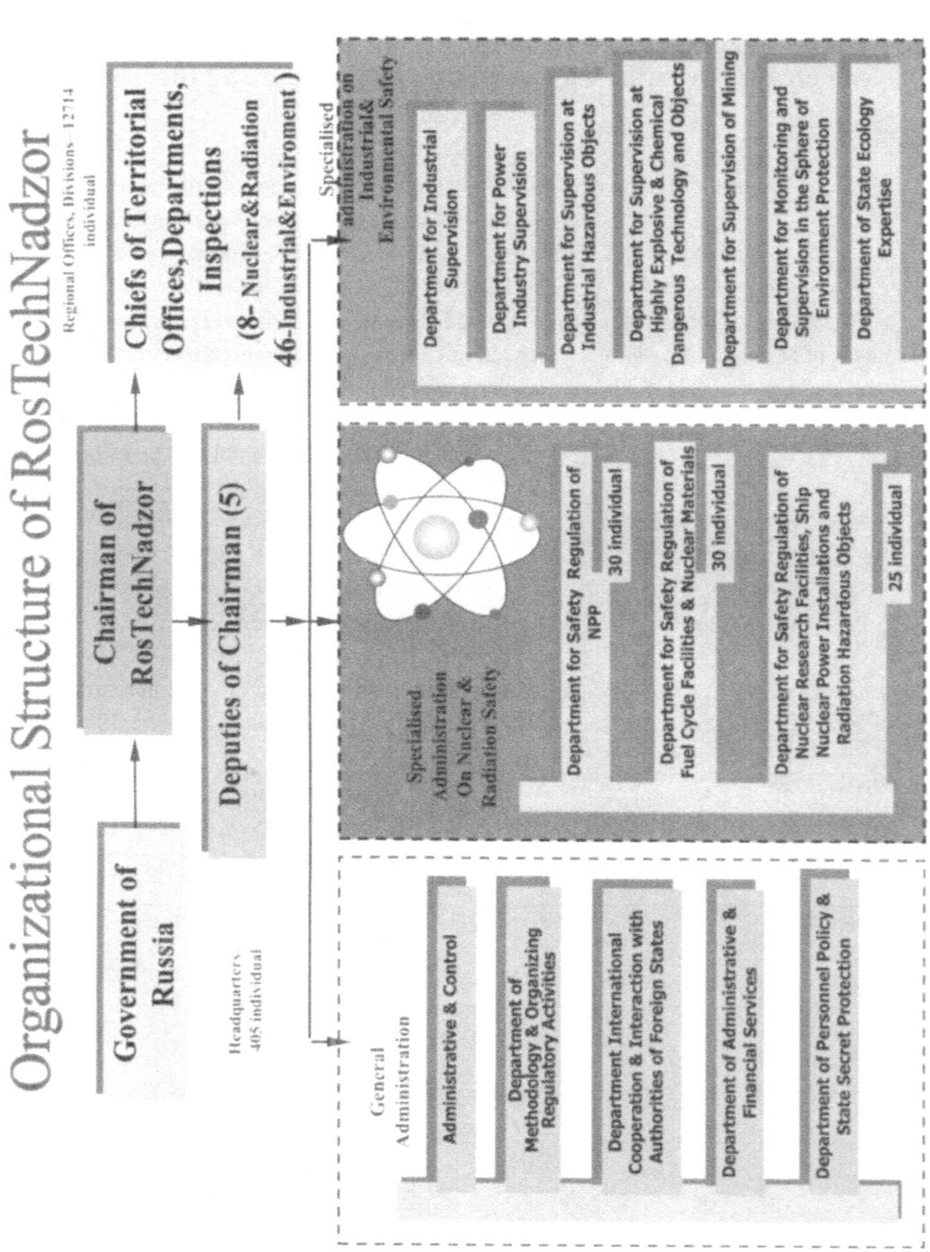
Organizational Structure of RosTechNadzor
Regional Offices, Divisions - 12714 individual
Government of Russia
Chairman of RosTechNadzor
Chiefs of Territorial Offices, Departments, Inspections (8- Nuclear&Radiation 46-Industrial&Enviroment)
Deputies of Chairman (5)
Headquarters 405 individual
General Administration
Administrative & Control
Department of Methodology & Organizing Regulatory Activities
Department International Cooperation & Interaction with Authorities of Foreign States
Department of Administrative & Financial Services
Department of Personnel Policy & State Secret Protection
Specialised Administration On Nuclear & Radiation Safety
Department for Safety Regulation of NPP
30 individual
Department for Safety Regulation of Fuel Cycle Facilities & Nuclear Materials
30 individual
Department for Safety Regulation of Nuclear Research Facilities, Ship Nuclear Power Installations and Radiation Hazardous Objects
25 individual
Specialised administration on Industrial& Environmental Safety
Department for Industrial Supervision
Department for Power Industry Supervision
Department for Supervision at Industrial Hazardous Objects
Department for Supervision at Highly Explosive & Chemical Dangerous Technology and Objects
Department for Supervision of Mining
Department for Monitoring and Supervision in the Sphere of Environment Protection
Department of State Ecology Expertise

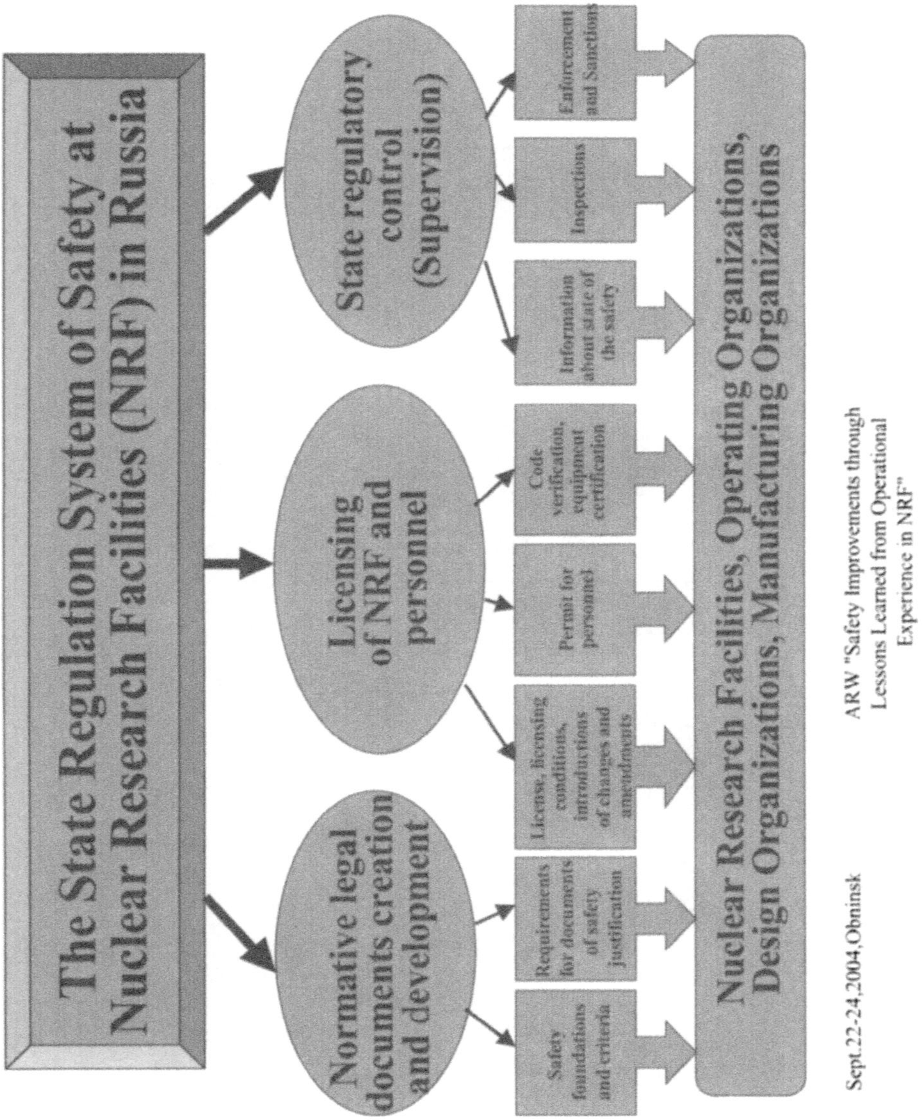
The State Regulation System of Safety at Nuclear Research Facilities (NRF) in Russia
Normative legal documents creation and development
Licensing of NRF and personnel
State regulatory control (Supervision)
Safety foundations and criteria
Requirements for documents of safety justification
License, licensing conditions, introductions of changes and amendments
Permit for personnel
Code verification, equipment certification
Information about state of the safety
Inspections
Enforcement and Sanctions
Nuclear Research Facilities, Operating Organizations, Design Organizations, Manufacturing Organizations
Sept.22-24,2004,Obninsk
ARW "Safety Improvements through Lessons Learned from Operational Experience in NRF"

The Russian Pyramid of Regulatory Documents for Nuclear & Radiation Safety and Authorities responsible for Documents

Constitution of Russian Federation (RF)	All-Russian referendum (nation-wide vote)
Federal Laws	State Duma, President of RF
Normative Legal Acts, International Agreements	President of RF (decree, orders), Government of RF (governmental regulations)
Normative Legal Documents (Federal Norms and Rules, Technical Regulations)	State Regulatory Authorities
Normative Documents, State Standards, Codes of Practice	Nuclear Authorities (Ministries&Departments)
Enterprise's Standards, Manuals, Programs, Test procedures	Operating & Special Organizations

Sept.22-24,2004,Obninsk

ARW "Safety Improvements through Lessons Learned from Operational Experience in NRF"

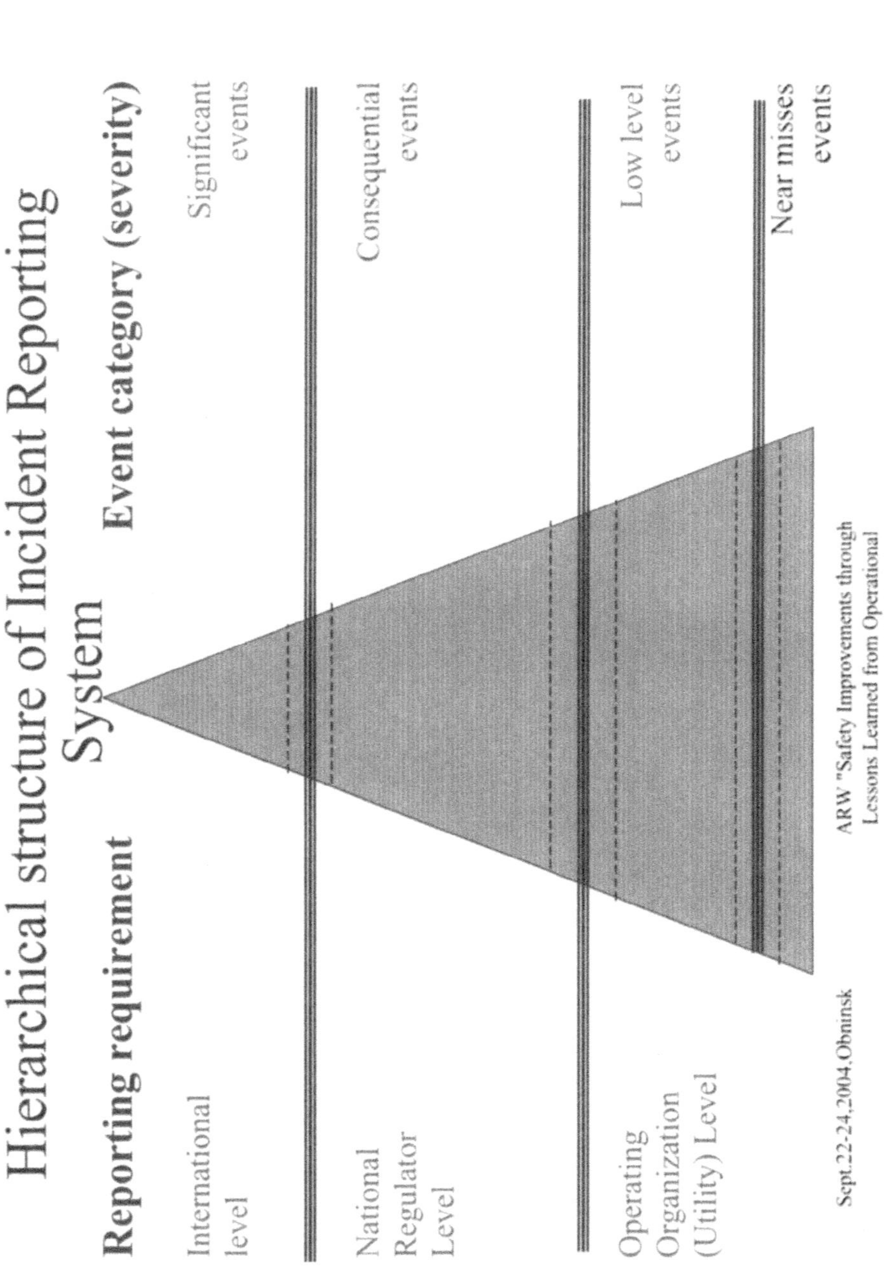
Hierarchical structure of Incident Reporting System
Reporting requirement
Event category (severity)
International level
Significant events
National Regulator Level
Consequential events
Operating Organization (Utility) Level
Low level events
Near misses events
Sept.22-24,2004,Obninsk
ARW "Safety Improvements through Lessons Learned from Operational Experience in NRF"

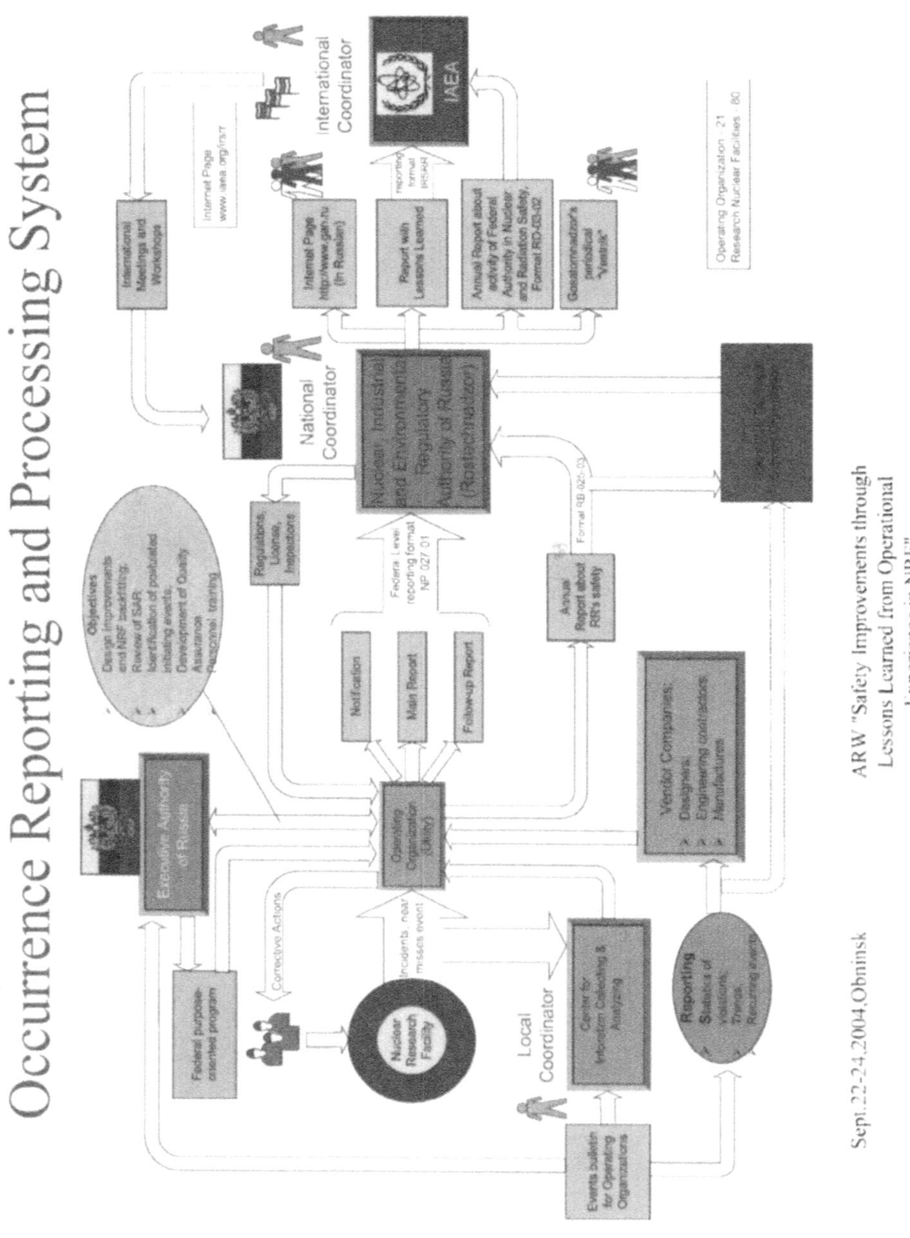
Occurrence Reporting and Processing System
Executive Authority of Russia
Federal purpose-oriented program
Objectives
Design improvements and NRF backfitting;
Review of SAR;
Identification of postulated initiating events;
Development of Quality Assurance
Personnel training
International Meetings and Workshops
Internet Page www.iaea.org/irsrr
Corrective Actions
Regulations, License, Inspections
National Coordinator
Internet Page http://www.gan.ru (in Russian)
International Coordinator
Nuclear Research Facility
Incidents, near misses event
Operating Organization (Utility)
Notification
Main Report
Follow-up Report
Federal Level reporting format NP 027 01
Nuclear, Industrial and Environmental Regulatory Authority of Russia (Rostechnadzor)
Report with Lessons Learned
reporting format IRSRR
IAEA
Annual Report about activity of Federal Authority in Nuclear and Radiation Safety, Format RD-03-02
Local Coordinator
Events bulletin for Operating Organizations
Center for Information Collecting & Analyzing
Annual Report about RR's safety
Format RB-025-03
Gosatomnadzor's periodical "Vestnik"
Vendor Companies:
Designers;
Engineering contractors;
Manufactures
Reporting
Statistics of violations;
Trends;
Recurring events
Operating Organization - 21
Research Nuclear Facilities - 80
Sept.22-24,2004,Obninsk
ARW "Safety Improvements through Lessons Learned from Operational Experience in NRF"

SAFETY IMPROVEMENTS THROUGH LESSONS LEARNED FROM OPERATIONAL EXPERIENCE IN NUCLEAR RESEARCH FACILITIES

French regulatory practices

David Krembel and Dorothée Conte
Directorate-general for nuclear safety and radiation protection, Sub-directorate for research facilities, dismantling of nuclear installations contaminated sites and radioactive waste, 10 route du Panorama, F-92266 Fontenay-aux-Roses, France

Abstract: The French nuclear safety authority (NSA) and its technical support (IRSN) started very early a process aimed at producing experience feedback to enrich initial safety analyses of nuclear research installations. The evolution of the regulatory framework dealing in France with experience feedback was very progressive. A technical and administrative organisation has been set up. Simultaneously, an important event database has been made available. NSA also plays a crucial role to liven up experience feedback between different operators. Finally, this paper presents the assessment and evaluation by NSA of the operational experience feedback from research nuclear facilities in France.

Key words: safety analyses, experience feedback, initiator, precursor event, event, incidents, accident, inspection, nuclear research facilities, INES scale, operator, regulatory provisions, nuclear safety, preventive actions, corrective actions, event database, transparency.

F. Lambert and Y. Volkov (eds.), Safety Improvements through Lessons Learned from Operational Experience in Nuclear Research Facilities, 111–126.

« Only the experience and reasoning coming out from our thoughts can educate us », (Alfred de Vigny)

1. INTRODUCTION

Nuclear facility operators and their competent authorities, must not only take particular care over risk minimisation that could affect the public and the environment. Both have to prevent and to reduce as low as reasonably possible their occurrence and consequences.

Experience feedback is a progressive safety improvement process based on the collecting and recording of data coming from nuclear facilities operations – in particular events and incidents – requiring the analysis, treatment and the diffusion of these data with the objective of making the facilities operations continuously more reliable.

Why and how are data coming from experience feedback of nuclear research installations collected in France ? Is there any specific regulation mandatory in this field ? What is the role of the nuclear safety authority of France (NSA) ? And how is all this information material shared between operators ?

This paper will first outline that experience feedback helps to enrich initial safety analysis of nuclear research facilities and will present the corresponding organisation set up in France. Later, we will show the assessment that is made from this experience feedback.

2. ENRICHING INITIAL SAFETY ANALYSES OF NUCLEAR RESEARCH INSTALLATIONS

2.1 Inherent limitations to initial safety analyses

In the process of setting up or modifying a nuclear research facility, a first step consists to get detailed safety analyses from the applicants, based on proven methods. The aim of such studies is to demonstrate that all risks presented by the planned facilities are under control. A set of regulatory provisions allow the Nuclear safety authority (NSA) to check the quality and exhaustivity of these analyses and consecutively to grant the requested authorisations.

However, national and international experience feedback are showing that, in spite of these risk analyses, a certain number of events or incidents occurs every year within facilities. Among them, some events could have been severe and present a risk to the public, or could have been considered as possible precursors of severe incidents.

Thus, aware of inherent limitations of initial safety analyses from the facilities, France started very early a process aimed at producing its experience feedback, not only concerning major events at national or international level, but also including minor events that could be considered as precursors of severe events, even if they didn't present any direct impact to the safety, the public or the environment.

2.2 The national organisation for collecting and sharing the experience feedback from research facilities

2.2.1 The progressive evolution of the regulatory framework dealing with experience feedback

- A national approach….

The main regulatory tool to fix mandatory provisions dealing with experience feedback in France is based on provisions of art. 5.III of the 11 December 1963 decree [1]. This decree defines a general frame of mandatory event notifications, applicable to all categories of nuclear installations, and in particular to research facilities. It prescribes that all accidents or incidents, nuclear or not, having or which could have serious consequences for the safety of nuclear facilities, must be notified immediately by the operators to the competent authority. Article 12 of the above mentioned decree sanctions operators having not notified incidents with penalties provided for contraventions of the 5th class (i.e. a maximum fine of 1500 Euros, 3000 Euros in case of repetition, or penalties restrictive or privative of rights). In practice nevertheless, such penalties have seldom been applied up this day.

Following the 1979 Three Mile Island accident, which notoriously showed the lack of precursor events detection, the French nuclear safety authority sent criteria to nuclear research facility operators, in case of events or incidents occurring within their installations. As such, operators have even the obligation to report events having only potentially small consequences on safety [2],[3]. For information, notification criteria applicable to the Atomic Energy Commission research facilities are provided in the appendix (see section 5). These provisions have been strengthened by the ministerial order and circular of 10 August 1984 prescribing general quality assurance and organisation rules in the French nuclear facilities [4]. This text addresses principally all events affecting the safety of installations called « deviations » or « anomalies », for which it is mandatory to have a system of experience feedback. In this respect, the 1984 quality ministerial order requires that detected deviations must be rigorously corrected and that preventive actions to stop a reoccurrence are conducted. This is in line with a systematic process of taking into account severe accidents precursors.

In the same way, concerning the diffusion of information regarding nuclear events or incidents, lessons learned from the 1986 Chernobyl accident lead France to set up, as of 1987, a severity scale for nuclear events, which the IAEA used as the basis for the INES scale (International Nuclear Event Scale)[5]. The purpose of INES is to cover events occurring in all nuclear facilities. For the Nuclear Safety Authority, INES facilitates selection of those of the many incidents and events occurring which are sufficiently significant for inclusion in its press releases :

1. All incidents rated at level 1 and above are systematically reported by the MAGNUC viewdata magazine and on the www.asn.gouv.fr web site. In addition, journalists are informed of incidents at level 2 and above by personally addressed press releases and telephone calls,

2. Level 0 incidents are not systematically made public by the NSA. The MAGNUC viewdata magazine and the web site report them in case of provisional classification, pending the results of complementary

investigations, if they are interesting in terms of safety analysis or methodology or if they are of particular interest to media (in case of release to the environment for example).

In other respects, it should also be mentioned that plant authorisation decrees, as well as general operating rules or ministerial technical provisions for each nuclear research facility reminds operators of their notification obligations of all kinds of events or incidents. These notification provisions have been recently highlighted again in a more generic way through the ministerial order of 31 December 1999 [6].

Further, since 2002 the NSA has started a responsibilisation process with the Atomic Energy Commission operators, allowing them, under certain conditions, to grant licences by themselves [7]. According to this process, the NSA asked the operators to inform it of the organisational and or technical experience feedback they gained from the internal licences, which present a significant interest in terms of lessons learned.

Finally, it should be mentioned that the general information obligation of the NSA should be consecrated next in the new and strengthened legislative framework that the French government has sent to Parliament [8].

- … with international links

France has taken part, as of 1994, in the Vienna IAEA Convention on Nuclear Safety [9]. The Convention states in its provision dealing with operations that incidents which are significant for safety must be notified in due time by the licence holder to the competent authority. This same provision stipulates that data collection and analysis programmes must be set up, that actions must follow, results obtained and conclusions drawn, and that existing mechanisms must be used to share information of importance with international organisations and with other operators and regulatory bodies.

Therefore, France has decided to include the case of research nuclear facilities in its second national report on the Vienna Convention implementation [10]; its third report to be published shortly will again address experience feedback issues in 2004. Additionally, France also shares its experience feedback at international level, through the IAEA « Incident reporting system for research reactors (IRSRR) » database, for events of particular interest.

2.2.2 The technical and administrative organisation set up by NSA

The consideration of experience feedback by NSA implies a good knowledge, follow up and timing of plants and their activities, in their material, as well as organisational and human aspects. To achieve this objective, NSA exerts a controlling action on the facilities, performed at the nearest level by its regional services. Issues with more generic aspects or specific difficulties, are considered at national level by the sub-directorate in charge of research facilities, among the central services of the NSA.

- Organisation at local level

The regional services of the NSA (Regional divisions for nuclear safety and radiation protection (DSNRs)) are immediately informed about events occurring within research facilities (by phone, fax...). As soon they have been informed, DSNRs may contact the operators to get additional useful information. They may also perform investigations during scheduled inspections or decide to organise reactive inspections for the most significative events. The DSNR reports to the central services of the NSA about results of their investigations. They also take part in NSA communication actions, and are in charge of the analysis of the event report sent by the operators in pursuance of notes [2] and [3]. Those reports which are the operator's responsibility to write, must consider the events chronology, initial causes, root causes, immediate and potential consequences and proposed corrective actions as well as an implementation schedule (see section 5 in appendix for the detailed content of event reports). Subsequently, DSNRs may check through inspections that operators have effectively put in place the announced corrective actions. In addition, DSNRs receive an annual plant safety report from the operators. This report highlights in particular key issues of the past year and presents a synthesis of quality activities with their associated experience feedback. All the information collected throughout the year about the plants operational experience are used by DSNRs to issue monographs for each installation. The DSNR's monographs, strictly for internal use, are updated on an annual basis, and allow the sharing of information and record keeping within the NSA of all key events that stand out as milestones in the plant's life.

- Organisation at national level

The central services of the NSA are notified from all the above-mentioned documents sent to DSNRs. Furthermore, a particular organisation has been set up to cope with incidents reported by the operators; it includes a specialised team whose task is to co-ordinate the NSA's actions within the sub-directorate in charge of research facilities.

In essence, this organisation isn't so very different from those set up to follow the experience feedback of power reactors. This incident-team updates the documentation applicable to events notification and treatment provisions applicable to research facilities. The incident team also updates an internal computerised event database. Its role consists in particular to seek to confirm the validity of the information sent by the operators (to check for adequate choices about events notifications criteria, quality of the event reports,...). It may be used for internal training sessions of the NSA concerning event management. Ultimately, it deals with events requiring experience feedback to other facilities in order to confirm or invalidate their generic character, i.e. affecting or likely to affect several facilities or operators. It ensures the adequacy of the corrective actions proposed by the operators.

2.2.3 An important event database also available to NSA at the Institute for radiation protection and nuclear safety (IRSN)

Since the early 1980s, NSA's initiatives have lead to the development of a database bringing together all reported events. This application is currently managed by the IRSN, NSA's technical support. This database which contained only a few reported events at its early beginning, nowadays grows by 800 new events each year, from which there are about 600 events for power reactors and 200 events for other facilities including research facilities. The IRSN brings also its expertise to the NSA for the technical investigations concerning the most safety-significant events and is able to conduct probabilistic studies in order to examine the influence of certain so-called « precursor » events, characterised by an occurrence probability above 10E–5 (for example, the possibility of a core melt severe accident).

2.2.4 To liven up experience feedback between operators: the crucial role of the NSA

One of the properties of research facilities comes from the often-unique character of each of them. Nevertheless, through its inspection's programme and through knowledge gained from the analysis of the lessons learned from operational experience of French or foreign facilities, the NSA could identify some generic subjects that may be of simultaneous interest for very different plants for which the NSA draws the operator's attention. This process doesn't only apply to research facilities; it embraces together all the French nuclear installations operators. For example, recently the NSA has published and circulated several letters to all operators dealing with experience feedback about unavailability management- and warning about a control desk ionising radiation event, about the prevention of atmospheric contamination risks associated with the presence of legionelosis bacteria due to the use of air cooling towers, and also, in the field of dismantling, about fire events initiated by cutting of cross wall metallic parts (pipes, beams,...), occurring outside the room where the cutting operations took place, because of the existence of thermal bridges (Figure 1). NSA's initiatives on those subjects can be limited to only drawing this to the attention of the operators. But NSA may also ask operators to send more in-depth analyses of such situations, to identify equipment in their plants potentially affected by similar failures and to propose corrective, or preventive actions.

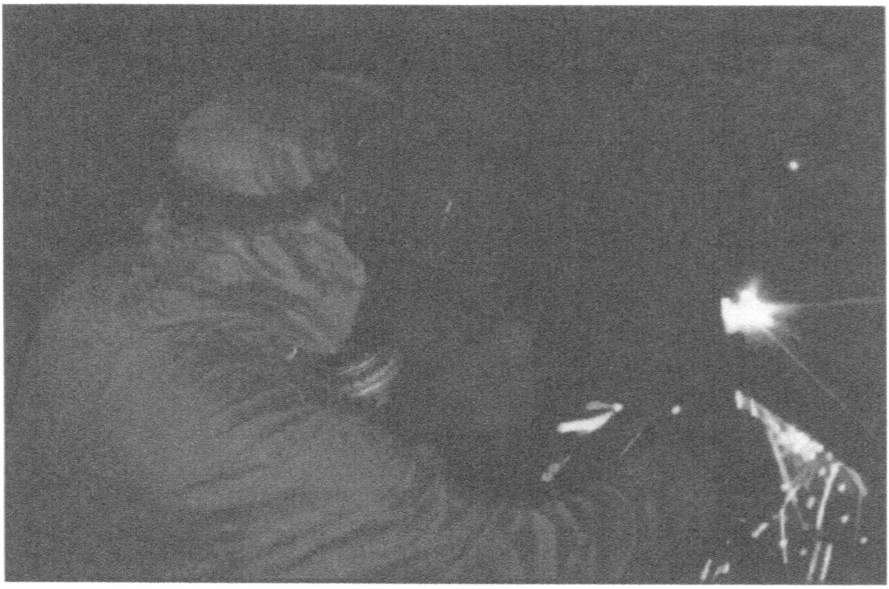

Figure 1 In the field of dismantling, NSA has recently drawn the attention of all operators about risks of fire events iniated by cutting of cross wall metallic parts (pipes, beams,...) occurring outside the room where the cutting operations took place, because of the existence of thermal bridges.

It should also be noted that NSA has recently decided to publish letters sent to operators on its Internet Website www.asn.gouv.fr following their plant inspections. Thus, with the aim of always seeking more transparency, those documents which are made available to the public, represents a very interesting source of experience feedback to all operators (Figure 2).

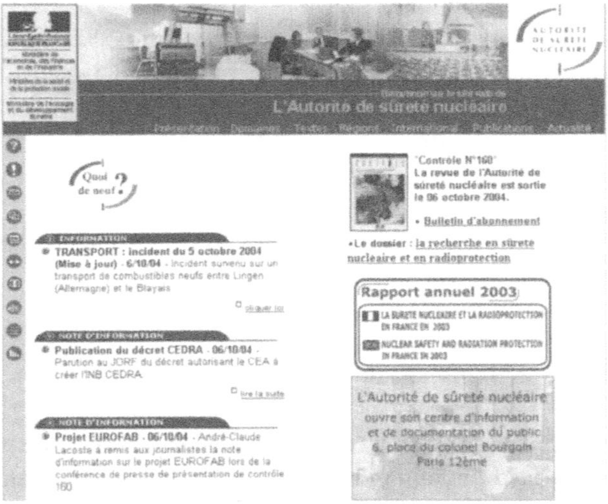

Figure 2 Results of inspections performed by NSA are regularly published on its website www.asn.gouv.fr and represent also a valuable source of experience feedback for all operators of research facilities.

Thus, from its central key position, the NSA plays a crucial role in the experience feedback's enlivening the whole process, by publishing and taking over the pertinent information at the right level and also acting with all nuclear facilities operators.

3. ASSESSMENT OF OPERATIONAL EXPERIENCE FEEDBACK AND ITS EVALUATION BY NSA

3.1 An assessment which essentially focusses the light on minor events

3.1.1 Number of incidents and classification on the INES scale

In total, 104 incidents have been declared in 2003 by 37 of the 71 research installations regulated in France by NSA. Between two thirds and three quarters of these incidents have been classified at level 0. Since 1998, no incident has been classified at level 2 or greater.

This result above all puts in the spot light these minor events which have no consequence for the public, the workers or the environment, but which in spite of all this, have been taken into account under the heading of operational experience feedback.

3.1.2 Principal installations concerned and initiators

The incidents declared concern, in the majority, the reactors of research and education, critical assemblies, laboratories, installations and facilities of industrial production. In particular, the assessment focuses on incidents linked to risks from loss of power supplies for certain plants. Besides this, human and organisational factors are the origin of almost three quarters of the events [11].

3.1.3 The absence of a correlation with the age of the installations

The average age of the research installations is relatively important. For all that, it does not seem very evident that a correlation of the number of incidents from the date the installations started operation can be established.

3.1.4 The increase in the number of incidents declared in recent times

NSA observes that since 1999 there has been a tendancy for the number of declared incidents to increase. In 2003, the rise has been 17 % as against the previous year. Between 1999 and 2001, this rise is explained by the proactive actions of NSA. Since then, several installations which used to declare few or no incidents in earlier years have declared more of them than usual, notably amongst those installations for the treatment storage of wastes, research reactors, and also certain installations which have been permanently shutdown, in the phase of shutdown or undergoing dismantlement. Finally, NSA also takes account, since summer 2000, of incidents in the area of radiation protection.

3.1.5 Improvement in the time allowed for declaration

On average over all the installations, NSA notes that the licensees generally conform to the declaration timescales prescribed by NSA in its notes [2] and [3] (see section 5 : declaration without delay for the serious incidents, under 24 hours in other cases), with a tendancy to improve. Deviations to declaration schedules fixed by NSA, when they occur, can be explained partly by the identification of incidents during inspections performed by NSA, after consultation of operators anomaly log books ; in effect, in certain cases the inspectors have observed that the licensees had not considered the necessity to declare certain anomalies recorded in their log books. The delay taken by the licensees to organise analyses to confirm the facts can equally explain this difference. However, these delays are purely qualitative indicators without any association with the quality of the documents sent to NSA.

3.2 The evaluation of the experience feedback by NSA

3.2.1 The event and incident identification process must be conducted with a positive state of mind

The operator should not consider that the number of events represents by itself a safety indicator. Nor should he consider that this event number is a reflection of the level of mistakes or of a poor performance, of its own organisation or in relation to the NSA. Such an outlook would be totally unproductive. A too repressive attitude resulting in the seeking to find scapegoats would be notably harmful to the level of transparency required in the experience feedback process. This process often consists of multiple causes identification which derive generally from a multitude of combined failures. NSA considers that the identification process must be conducted with a positive state of mind: one should consider the event or incident identification as a commitment to enriching experience feedback and improving the overall process. The importance of this process shouldn't be underestimated.

3.2.2 NSA is looking for improvements regarding the content of event reports

The analysis by the NSA of event reports shows that the section dealing with potential consequences of assumed deteriorating situations is sometimes insufficiently developed. Furthermore, NSA has noted a certain reluctance from the operators to tackle the human and organisational factors in the event analysis. Of course, making the responsible persons feel guilty has no place in this analytical process.

3.2.3 Lessons learned from operational experience have put initial safety analysis in question

Finally, NSA has noted that lessons learned from about ten incidents a year have put in question assumptions made during initial safety analyses or have shown an insufficient consideration of risks at the design stage of the facilities [11]. In these cases, the plant safety analyses have been updated (Figure 3) or additional risks studies have been started.

Figure 3 Experience feedback is an important part of the safety reviews of old nuclear research installations considered by NSA's advisory committees.

4. CONCLUSION

France aims to continuously seek improvements in the safety of research nuclear facilities through consideration of their experience feedback. There are many and complementary sources to establish this experience feedback: through nuclear event or incident declaration reports, periodic operation information reports, actions following inspections, etc. This experience feedback is inter alia turned to NSA's advantage to keep the memory of key facts/events occurring during the plants life, for training purpose of its personnel, and is very effectively used during periodic plant safety re-assessments.

The experience feedback process in France was set up very progressively, as and when operators acquired a greater radiation protection and safety culture. This long-term process was strongly initiated by NSA and its technical support, especially towards taking into account the lessons learned from small events or deviations. Finally, through the values which personifies its action: transparency, independence, proficiency of its personnel, as well as the strict application of its operation of experience feedback, the NSA also actively contributes to give confidence among the public in the safety of nuclear research facilities.

5. APPENDIX

Declaration to the NSA of significant accidents or incidents pertaining to the safety of nuclear facilities operated by the Atomic Energy Commission (see [2])

5.1 Definition of accidents or incidents which should be notified

1. Transfers of dangerous goods – radioactive or not – leading to a situation where the level of safety of the plant is reduced or to important risks for workers.
2. Accidents or incidents, whatever is their severity, as soon as it appears that they could lead to erroneous interpretation or malevolence in the media or for the public.
3. Malicious incidents that could threaten the plant safety.
4. Aggressions due, either to natural phenomena or to human activities, having actually or potentially and in a significative way, affected the plant safety.

5. Accidents or incidents of nuclear origin or not, at the origin of loss of life or severe injury requiring inter alia evacuation of the injured people to a hospital centre.
6. Significative accidents or incidents having lead to the loss of functions from all lines of defence set up between the dangerous goods and the persons and at the origin of a dispersion of these dangerous goods [...].
7. Accidents or incidents which, although having not breached the totality of lines of defence , lead to, or could have lead to a significant dispersion of dangerous materials or to a significant individual radiation exposure, on or off the facility.
8. Faults, degradations or failures having affected an essential safety function, which had or could have had significant consequences, whether they have been identified during or outside normal operation. This is particularly the case for faults, degradations or significant failures affecting one of the lines of defence, one of the safety systems associated with the lines of defence or one of the protection or emergency systems such as power supplies.
9. Incidents having lead to the bypassing of one or several security limits as defined in the plant's technical provisions or to a common mode failure of important safety systems.
10. Even minor events, affecting an important safety function which present a repetitive character for which the cause has not been identified or which may be precursors of accidents.

5.2 Accidents or incidents declaration process to the NSA

A declaration must be sent without delay for:

1. Accidents having lead to loss of life, important exposure to ionising radiation or severe injuries ;
2. Events having lead to significant and unexpected release of dangerous materials offsite the plant;

3. Case where the operator considers opportune to inform quickly the competent authority – of an incident or accident, which could lead to erroneous or malevolent interpretations for example-, the event seemingly not requiring to put in action the internal emergency plan.

For other accidents or incidents, the declaration must be performed in the 24 hours following their identification. Those declarations should be performed in the same time allowed even in the case of absence of the first results of investigations conducted to find out the circumstances of the event.

Independently from the declaration, a detailed event report must be sent within the one-month period following the date of the significant accident or incident. The event report should inter alia mention:

- The event description (circumstances, chronologies,...);
- Its causes ;
- Its consequences, in particular regarding radiological aspects ;
- Its analysis ;
- Remediation actions taken to avoid its renewal and the associated experience feedback.

References

[1] Decree n° 63-1228 of 11 December 1963, as modified, pertaining to nuclear facilities (Official Journal of the French Republic of 14 December 1963)

[2] Letter SIN n° 480/83 of 25 January 1983 pertaining to the accident or significative incident events concerning nuclear research facilities from the French Atomic Energy Commission

[3] Letter SIN n° 5651/84 of 10 October 1984 pertaining to the accident or significative incident events concerning nuclear research facilities from other operators

[4] Ministerial order and circular letter from 10 August 1984 pertaining to design, manufacture and operations of nuclear installations (Official Journal of the French Republic of 22 September 1984)

[5] International nuclear event scale – 2001 user's guide edition – issued jointly by the International atomic energy agency and the OECD agency for nuclear energy

[6] Ministerial order of 31 December 1999 addressing general technical regulation to prevent harmful effects, and external risks resulting from the operation of nuclear facilities (Official Journal of the French Republic of 15 February 2000)

[7] Note SD3-CEA-01 of 16 May 2002 pertaining to the supervision by NSA of Atomic Energy Commission's internal licences

[8] Draft law on nuclear security and transparency

[9] Vienna Convention on Nuclear Safety signed by France on 20 September 1994

[10] Second report of France on the implementation of the Convention on Nuclear Safety (September 2001)
[11] Safety assessment of the Atomic Energy Commission's nuclear facilities from 1999 to 2001 (November 2002)
[12] Nuclear safety and radiation protection in France in 2003.

THE RESPONSIBILITY OF HIGHER MANAGEMENT WITH RESPECT TO THE SAFETY POLICY OF RESEARCH CENTRES

Paul Govaerts
SCK•CEN, Boeretang 200, B-2400 Mol, Belgium

Abstract: The management has to implement and to preserve a high level of safety culture. The environment of a nuclear research centre induces specific challenges to this task. Those aspects relate to the nature of the installations, the human factors in a scientific environment and the specificity of the nuclear hazards. The rise and decline of safety cultures will be discussed. A sustainable safety culture requires safe installations, a convincing commitment by management, a clear assignment of responsibilities and the implementation and control of feasible procedures. All issues related to communication are very important. They require a specific social climate where blame-free reporting is essential. This climate has to be supported by an active participation of all levels of the organisation. The main threats to safety culture are over-confidence and the denial of small incidents to preserve the image of safety.

Key words: nuclear safety; safety culture; management; research centre; human error; radiation protection; training; communication; feedback of experience.

1. SPECIFIC SAFETY ASPECTS OF A NUCLEAR RESEARCH CENTRE

1.1 The installations

As it is the case at most complex industrial facilities, nuclear research centres present a broad spectrum of risks: Fire hazards, mechanical hazards e.g. at the decommissioning workshop and due to the manipulation of heavy loads, chemical hazards at the laboratories and by the use of liquid metals, electrical hazards, due to the use of heavy motors and heating elements, mining hazards in the underground research laboratory and, last but not least,

F. Lambert and Y. Volkov (eds.), Safety Improvements through Lessons Learned from Operational Experience in Nuclear Research Facilities, 127–136.

the nuclear-related risks: irradiation or contamination of workers, release to the environment and nuclear excursions due to reactivity accidents.

Experiments are mostly set up for a rather short period within an ageing host facility that can accept a series of successive experiments over many years.

This situation is rather specific from the point of view of safety: the experiments have often to cope with childhood diseases, while the host facility has to cope with ageing problems. Figure 1 shows the typical bathtub-curve of reliability versus age. For "young" experiments, the management has to decide whether the starting problems are indicative for the poor quality of the installations or whether the initial failure rate will quickly decrease to a normal level after the elimination of a few artefacts. For the hosting installations presenting an increasing rate of failures, a timely decision about a refurbishment or end-of-life with or without replacement has to be taken into account.

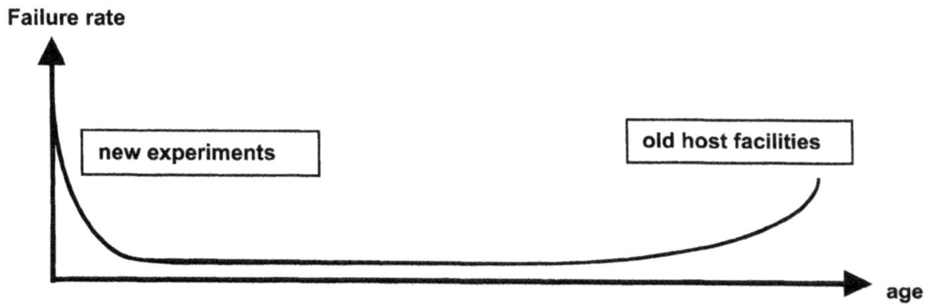

Figure 1. Research centres have to cope as well with childhood diseases as with ageing problems.

Since some experiments have a very short lifetime (days, months), it is often considered as not justified to install expensive engineered safety features. In this way, the protection at the source shifts to a focus on personal protection equipments.

Experiments require often transient phases presenting a higher risk than routine operations. Frequent modifications require continuous training and a great awareness to adapt the technical specifications, procedures and inspections to the up-to-date status of the installations.

Although some activities know their contractual constraints, the economic pressure is less important at a research facility than for an industrial production unit as e.g. a nuclear power plant. The conflict between safety and availability will often turn out in a victory of safety. This context also allows avoiding as much as possible maintenance during the operation, a situation that often leads to initiating events of notional incidents, such as the Three-Miles-Island accident.

1.2 The human factor

The middle management of a research centre is mostly scientifically oriented. This scientific spirit could lead to a risk taking attitude and a reluctance towards excessive regulations, procedures and external control. Since research budgets are always exhausted, there is also a real risk that preference is given to new investments and new scientists, reducing the technical manpower to a minimum. These kinds of risk are the most important for smaller experimental set-ups. Large facilities are organised and managed as common industrial installations.

The individualistic orientation of the scientists requires a strong safety service that will be considered by the scientists as a real support. The safety services may however not exempt the managers and other actors of their responsibility and accountability.

1.3 The nuclear factor

Nuclear safety presents some specificity that has an impact on the attitude of the public, workers, authorities and ipso facto of the responsible management. The main specific issues are related to the non-familiarity with radiation, the non-zero risk of very low exposure, the possibility of severe accidents and the production of radioactive waste.

Although we are living daily with radiation coming from natural and medical sources, nuclear installations are often still considered as mysterious black boxes, where an untouchable elite is playing with the health of people. The non-familiar nature of radiation hazards increases the perceived risk. Since the complexity and inaccessibility of most installations, this black box concept could also be the image as seen by workers.

Since the exposure to ionizing radiation is assumed to present some risk without any threshold, it is difficult to define safe situations. Each exposure or release has to be assessed in the context of the optimization of protection. The residual risk cannot be considered as trivial. This makes the communication to the public and to the workers often difficult.

The possibility of severe accidents with a very low probability but entailing important health and economic consequences requires of course a permanent awareness by all actors and in particular by the management. Last but not least, the production of radioactive waste has some specific impact on the management of nuclear installations. A strict policy of the physical control and inventarisation of waste is required, as well as a budgetary policy considering future, uncertain, costs.

The hazards of radiation and of potential accidents are invisible. People correct their behaviour commonly in direct contact with visible hazards, such as e.g. fire. In this way, it is important to make the radiological hazards as visible as possible at the work place e.g. by redundant and diversified monitors with observable alerts and personal alarm dose- and doserate meters.

2. STRENGTHENING OF THE SAFETY CULTURE

2.1 The rise and fall of safety culture

Figure 2 shows a schematic view of the "natural" evolution of the safety culture. It is the management's main task to keep the level at the top by coping with natural tendencies leading to a decline. The next paragraphs will describe the conditions of success and the mechanisms of decline in more detail.

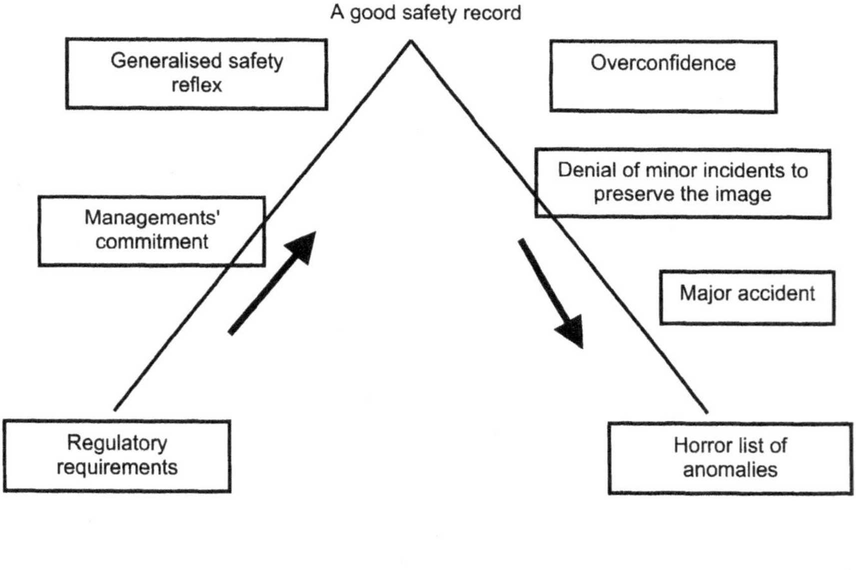

Figure 2. The rise and fall of safety culture.

2.2 The regulatory requirements

The first condition to safety is the excellent quality of the installations. In a first phase, the quality will be imposed by regulatory requirements ensuring the conventional ingredients of a defence-in-depth approach. It is important also to set up adequate reguirements for experimental devices. At SCK•CEN experiments have to pass a three-stage licensing system:

1. Is the experiment feasible without endangering the general safety of the host facility?
2. Is the design of the experiment adequate?
3. Are the inspection programmes and the procedures adequate?

For major experiments (e.g. for each device in a research reactor), this licensing process is surveyed by external safety authorities.

2.3 A convincing commitment to safety

In those days of quality assurance systems, it is fashionable that the management includes formally safety objectives in a written mission statement. This statement should encourage the personnel to behave in a way the objectives will be realised. Such a statement is however only convincing under strict conditions:

- the commitment must be materialised by reserving a reasonable budget for safety (personnel and equipment);
- managers must show the good example by their behaviour on the work place and by their competence on safety issues;
- managers should respond immediately to each indication of a lack of safety;
- the management should support a zero tolerance policy against bad intentions or unacceptable anomalies, but have to be tolerant to human error in order to keep open communication feasible.

Some of these conditions can be threatened by the above-mentioned specific nature of scientists.

2.4 The generalisation of the safety reflex in the organisation

2.4.1 The management has to offer the substrate

A generalised safety reflex can only become sustainable within a suitable environment, created and preserved by management. This substrate consists of a clear structure, adequate training, an efficient feedback of experience and a social climate allowing open communication.

The organisation chart must assign clearly the responsibilities. It is important to put the accountability on those who control the activities and the resources, without minimizing the individual responsibility of each actor. The interaction with safety services has to be well understood. Safety services are at the sideline, surveying and supporting, and they are accountable in this role. Other challenges for the top are the establishment of procedures and the control of their application and the support of a communication culture.

2.4.2 The procedures

It is important that the operators can dispose of detailed guidelines for their behaviour in all foreseeable normal and abnormal situations. The drafting process of a procedure has to be started by the management specifying the technical objectives and analysing the safety aspects. It is crucial to indicate the limits of the applicability of each procedure. At the limit, the operator has to contact a higher level in the organisation. This higher level has to accept that the operator revealed his limit of insight or power of control. The management has also to explain to the operators the "whys" of the procedure, indicating the objectives, the safety risks and the implicit multi-barrier protection. This will convince the operators not to short-circuit some apparent useless steps in the procedure. It is recommended to ask the operators themselves to draft the operational procedures, taking into account the feasibility in practice. The management has to establish a control system to survey whether the procedure is applied and is practically feasible, is meeting the objectives and preserves safety. The safety services have to survey the whole process.

2.4.3 Training and feedback of experience

Each level of the organisation has to be trained on the basics of safety and radiation protection. It is important that higher management is well aware of all aspects. Specific training has to be given to operators of specific installations or to those who have to execute specific interventions. This training can be supported by computer simulations or by tests on "cold" (non radio-active) physical mock-ups. The training programme must explain the rationality of the procedures and refer to the feedback of experience of the concerned installation and of similar activities worldwide. On the other hand, the lessons learned during training sessions have to be used to polish the procedures.

2.4.4 An open communication

It is important to realise a real open communication on safety problems through the organisation. This requires the insurance of blame-free reporting. Bad consequences due to human errors may not be sanctioned, unless they are caused by an unacceptable behaviour of the operator. It must also be acceptable that an operator reports an error by his friend, without empoisoning the social climate. Each anomaly must be communicated to all concerned, by preference together with the results of the analysis and the corrective actions. For the management, it is important to avoid the natural denial reflex, where small incidents are hidden to preserve the image of safety and good management. In this context, a common approach to safety culture between utilities and authorities is needed. Often the management is enforced to blame its personnel under pressure of the authorities or the media.

2.5 Overconfidence

A good safety record and a defensive position against outside criticism can lead to the conviction amongst the management and the operators that the safety of the installations is undoubtful, a law of nature. By overconfidence operators will adapt gradually their behaviour to the easiest way. They will short-circuit steps in the procedure that are felt as time-consuming and do not add to the technical efficiency. A speaking example is the Tokaimura accident, where safety rules were not respected most probably by the lack of knowledge about criticality risks, by the lack of managerial survey and by overconfidence. It is the task of the management to recall periodically that safety is not a law of nature but can only be ensured by a never ending effort.

2.6 Preservation of the image

Each organisation wants to preserve an image of excellence to the outside world. This attitude can become dangerous when it is reflected to the internal communication on safety problems. Organisations with a good safety record often refuse to consider smaller anomalies as serious. The tolerance of those anomalies is an invitation to additional problems. The reaction of the safety authorities and the media is crucial in this process. An overreaction to small reported problems will refrain organisations to report new events. This mechanism seems to play a role in Japan, where e.g. negative inspection results were hidden to the authorities to avoid an excessive reaction.

2.7 The horror list of anomalies

The independent examination after a severe accident reveals often a long unbelievable list of anomalies. The most common issues are:

- the denial of precursors;
- the lack of inspections;
- the lack of maintenance;
- the ridiculisation of whistle blowers;
- the delay of corrective actions;
- ....

The management of research installations has to look at those lists, with the intent to improve their own safety culture.

3. THE MANAGERIAL CONTROL SYSTEM

The safety of SCK•CEN is controlled by three parallel systems:

- the legal safety structures:
- the social survey system;
- the management survey system.

Those systems are not independent. It is obvious that all hierarchical levels and the legal internal safety services play an important role in each of those.

The legal safety structure foresees three levels:

- the internal safety services (industrial safety, physical protection against the risks of ionizing radiation and medical survey); Those services report directly to the general manager, they have to authorize specific installations and interventions, they survey hazardous operations in real time, they check the quality of protection features and monitoring, they are in charge of the safety training and have to organize the emergency management. The three services are integrated and concertation between individual experts is stimulated.
- the regulatory body: an external independent technical support organisation that has to authorize specific installations and interventions, and has the duty to survey the quality of the internal safety services. Representatives of the regulator have free access to the installations and may discuss safety issues with the operators.
- the safety authorities: prepare the licences for the competent minister and are to be informed about doses, releases and in case of legally specified anomalies. They have the right to suspend the licence or to impose specific working or reporting conditions.

The social survey system is based on the daily control by representatives of the trade unions. They are often used as bypass of the hierarchy and have an easy access to the highest level of the management. As legally required, a safety review meeting with elected representatives of the trade unions, the safety services and the management is organised on a monthly basis. This committee agrees annually on a generic action plan and can enforce constraints below the legal limits, as it is e.g. the case for the individual dose burden.

The management survey system is based on a quality assurance system, implying a set of procedures and instructions that are often related to safety. The most important is the ALARA-procedure, requiring a risk assessment and a prediction of consequences (e.g. individual and collective doses) before each hazardous operation. The depth of the analysis and the review procedure depend on the perceived severity of the operation. The procedure foresees also a formal feedback, comparing the predictions with the observed consequences. The most important cases are discussed within the ALARA-committee, composed by members of the hierarchy, independent experts and members of the safety services. Specific committees are created to review the experiments in the BR2-reactor and in the underground laboratory. Each committee is reporting to the general manager, who can be invited to specific meetings. The safety aspects of the main research reactor, BR2, are reviewed annually by an international safety review committee, chaired by the general manager of SCK•CEN.

The management is also informed by the evolution of safety by a keyboard of safety indices, resuming on a monthly basis, the distribution of individual doses, the collective doses, the atmospheric releases, the amounts of radioactive waste produced and evacuated, the number of accidents, the working days lost and the number of real and false fire alarms. This keyboard is a useful tool to guide the safety culture of our facilities.

At nuclear installations, the accumulation of radioactive waste could become a significant threat for the quality of working conditions. The management has to imply a strict system to collect, inventorise and evacuate the waste production. Adequate monetary provisions have to be constituted to handle the waste coming from the future dismantling of today's installations and experiments.

The highest management level has to guarantee the quality of the emergency facilities, procedures and know how. The general manager is called to co-ordinate the emergency plan in case of an accident. This requires a personal comprehensive insight in the risks of the installations, allowing an efficient concertation with technical and safety experts in case of a crisis.

The external communication is an important aspect of the emergency management. The management has to survey the communication with the safety authorities and with the press media in order to facilitate the protection of workers and members of the public.

4. CONCLUSIONS

The nature of the facilities of a nuclear research centre, the economic situation and the characteristics of the personnel have a specific impact on safety issues. The management has to keep a real safety culture alive. This requires a formal organization of proactive risk assessments, quality control of installations, quality assurance of procedures and training at each level.

It is important to create a spirit of open communication to allow an efficient feedback of problems. Safety control has to be redundant in a way the bad news reaches always the management level. A specific care is needed for emergency preparedness and external communication.

A good safety record is not perennial. Some natural laws of human behaviour lead to the decline of safety culture. Overconfidence, considering safety as a law of nature and the denial of minor incidences to preserve the image of the organisation undermine the safety behaviour. Managers must be aware of these phenomena and must counteract complacency by recalling that not safety but risk is a law of nature.

In a nutshell: the manager has to take the difficult decisions, has to detect the top of the iceberg and to enforce some less popular procedures. This requires safety-minded managers.

ORGANIZATION AND METHODS USED BY THE CEA SACLAY CENTRE TO IMPROVE OPERATING PROCEDURES AND PROMOTE BEST PRACTICES IN NUCLEAR RESEARCH FACILITIES

B. Sevestre , P. Reynard, G. Cadolle & P. Riband

Commissariat à l'Energie Atomique, centre de SACLAY

Abstract: Since 1952, the Saclay CEA Center has been a leading player in the rapid development of nuclear science and technology in France. It hosts a wide range of activities, from basic to applied research in a wide range of fields and disciplines, including physics, metallurgy, electronics, biology, climatology, simulation, chemistry and environmental science.

These activities involve a complex pattern of partnerships with various research and academic institutions, both public and private, and with industrial and service providers. Seven thousand people, including 4,000 CEA workers, use and manage hundreds of laboratories and support facilities, including 9 major nuclear facilities (4 reactors, 3 laboratories or production facilities, 2 waste management facilities), many smaller devices involving radiological management, and large-scale dismantling activities.

In-service experience is based on two aspects :

- identification and correct treatment of each event related to safety, to improve operating procedures,

F. Lambert and Y. Volkov (eds.), Safety Improvements through Lessons Learned from Operational Experience in Nuclear Research Facilities, 137–185.

- organization of efficient exchange between peers in charge of nuclear facilities, to promote best practices.

  The CEA Saclay organization and methods concerning operating experience feedback will be presented, and illustrated by several examples:

- two examples of safety-related events in a research reactor,

- two examples of generic analysis of recurrent events or risk factors.

1. INTRODUCTION

1.1. Research activities (see Appendix 1)

The CEA Saclay Center is involved in a very wide range of research activities that make extensive use of industrial or scientific cooperation.

Further details about these activities are given in appendix 1.

Research in the field of nuclear energy

The Center is specialized in computer simulation focusing on two main areas:

- reactor operation,
- behavior of materials and structures.

Development work in these fields uses large-scale experimental facilities devoted, in particular, to the qualification of new nuclear fuels and behavioral studies for structural materials exposed to neutron radiation (e.g. OSIRIS reactor and the LECI laboratory). (see appendix , picture 1)

Technological research

The Center's chief activities are concerned with developing embedded software and innovative materials, as well as performing radioactive measurements (in particular for national standards at the *Laboratoire National Henri Becquerel*).

Physical sciences

The Center's activities in this area include particle physics, nuclear physics and astrophysics, together with the study of matter down to the atomic and molecular scale. Collaborative work is particularly well developed in this area, which demands large-scale scientific instruments – some of which are located on the Saclay site (ORPHEE) – or directly involve the Center's teams (the GANIL facility in Caen).

Life sciences

At CEA/Saclay, research does not only focus on the effects of radiation on living organisms, but also on how ionizing radiation or other non-invasive techniques can be used in observation instruments to help us to learn more about the various functions of some organs , (cardiology, cognitive functions, etc.). A major development in this area is NEUROSPIN, a new piece of equipment soon to be built at the Center, using technologies developed by specialists in physical sciences.

Climate and environmental sciences

The Center's involvement in this field of research sprang from the conjunction of its capabilities and know-how in computer modeling and its high-precision measuring equipment, designed for plotting radioactivity and chemical elements to date and characterize ice or sediment samples.

1.2. Organization, facilities and equipments

1.2.1 Organization and people involved

Seven thousand people work at the Saclay Center, including some 4,000 CEA employees.

Non CEA people work on the site under various conditions:

- some are under contracts signed with outside service providers for work involving modification, maintenance, inspection, periodic testing or equipment operation,

- others are researchers working at the Center for joint laboratories (CEA/CNRS or CEA/Universities), or for temporary cooperation projects, training courses, and so on.

In the same way, CEA/Saclay employees can work off-site as part of joint projects conducted in France or in an international context.

Whatever the case, contracts or agreements must clearly define responsibilities and rules related to security and safety.

As a general rule:

- As operator of the Center's facilities, CEA must assume responsibility for all matters relating to security and safety.

- In addition, all employers are responsible for the security and safety of their own personnel, with particular regard to radiological protection, although CEA is also responsible for ensuring that all those working on its sites comply with labor regulations and health and safety rules.

Within its internal organizational structure, CEA entrusts these responsibilities to the Director of each center (including CEA/Saclay), who heads the following organizational structure:

- An operational line organization composed of a chain of command, running from the director of the center through the various heads of department and down to each facility manager. Considerable powers are delegated to the facility managers, giving them the authority to ensure the security and safety of all those working in their facility. In the same way, each CEA center reports to an operational manager who, in turn, reports to the CEA Chairman.
- A staff organization built around services specialized in every area of security and safety: occupational safety, environmental protection, radiological protection, health service, security forces, and the nuclear safety unit. In particular, the nuclear safety unit is responsible for inspection, which consists of performing spot checks to ensure that the technical organization rules relating to safety are observed and properly adapted. Following the same principle, DPSN (the nuclear safety and radiological protection division) pools the technical and regulatory expertise at the CEA central level.

The center's quality system is the ideal tool for describing and promoting the efficiency of this organizational structure, drawing on lessons learned from deviations, incidents, audits, stakeholders' opinions, measurements, and indicators.

Regarding organizational aspects not called upon in routine situations (e.g. special interventions and crisis organization), exercises are planned on a regular basis (every year, at least one exercise per facility and at least one exercise for the center as a whole). Each exercise generates fresh information.

1.2.2 Nuclear facilities and equipment

The Saclay CEA Center has a very diversified park of facilities and nuclear or radiation-emitting equipment governed by several types of administrative authorization procedures according to their importance with respect to the personnel, the public and the environment..

- 4,000 radioactive sources,
- 60 X-ray and crystallography generators,
- 55 ICPE [facilities classified for the protection of the environments, with respect to radioactive risks)
- 11 Basic Nuclear Installations.

The following part of this presentation concentrates on the basic nuclear installations and associated lessons learned. At the Saclay Center, this involves three reactors, three laboratories or production facilities, two waste processing facilities and other facilities currently being decommissioned.

These facilities are characterized by major safety issues: in the case of an hypothetical, uncontrolled accident, they could be the source of considerable impact on the public and the environment.

Control of their safety is based on the principles of defense in depth:

- robust design and operating procedures to prevent incidents.
- technical provisions and procedures to detect incidents and prevent them from leading to an accident,
- theoretical analysis of a certain number of accidents and predefined measures to limit consequences and return to a safe state.

According to and in proportion to this safety issue, a certain number of lines of defense are therefore defined and periodically tested if they are not used in routine operation.

A well-designed defense in depth system allows the possibility of some failures, that won't lead to further consequences .It is therefore essential to identify and analyze all deviations, incidents or events that reveal insufficiency or degradation of the defense in depth system, to ensure continuous evaluation of the completeness and solidity of the defense in depth system,

This is achieved basically in two ways: periodic safety reviews and operating experience feedback.

- Every ten years, a safety review is conducted in each facility, consisting of a safety analysis that takes into account, first, changes in research programs or activities carried out in the facility; second, changes in safety regulations and practices; and finally, the operating experience of the facility in question and other similar facilities. At the end of the review, decisions are made, based on technical and economic analyses, to bring the facility to the highest level of safety and radiological protection that can reasonably be expected.

- On a continuous basis, deviations or operating incidents are identified, classified and processed according to the procedures of operating experience feedback described below.

2. OPERATING EXPERIENCE FEEDBACK

2.1 Identification and processing of each safety-related event

Detection and processing of deviations

Basic Nuclear Installations (INB) have a quality organization system that specifies how deviations are to be detected and processed.

Deviations are usually detected by operating personnel during control or maintenance activities or during internal inspections relevant to these activities. Deviations may concern not only non-compliance with a technical parameter or procedure, but also an event considered important by the operating team.

Declaration of significant safety-related events

The Nuclear Safety Authority has defined the events considered significant for safety, which must be declared rapidly, followed by a detailed report.

In CEA's organization, this requirement is the responsibility of the Center's Director. To meet this requirement, CEA/Saclay director asks that the facility manager of each individual INB advise it immediately of any deviation likely to correspond to the Nuclear Safety Authority's declaration criteria. The Identification of these deviations calls on the better judgment of the manager and safety experts. The declarable nature of the event is examined during a meeting chaired by the Center's safety unit, the operating line and, depending on the nature of the event, the Center's radiological protection department or any other technically competent unit. If the event proves to be declarable, a declaration message is proposed to the Center Director. Otherwise, a report listing the reasons for non-declaration is established, archived and kept available for Nuclear Safety Authority inspectors. The event declaration indicates its classification on the international nuclear event scale (INES). This classification is decided by CEA/Saclay management following consultation with CEA's Nuclear Safety and Radiological Protection Division (DPSN). Any event with a level equal to or greater than 1 generates a press release prepared by CEA/Saclay management's Communication and Public Affairs Department after consulting with the safety unit.

Analysis of significant safety-related events

Significant safety-related events are analyzed by the Basic Nuclear Installation manager, and the results are formalized in a report. This document, which is checked by the Center's safety unit, serves as the detailed report requested by the Nuclear Safety Authority.

The analysis included in this report must define measures to be taken in the facility to prevent similar or more severe events. In addition, preparation and distribution of the report must allow other facilities to take advantage of experience feedback from the event.

Analysis of significant safety-related events is carried out according to a method recommended by CEA, which emphasizes:

- collection of data on the principal factors of the incident and study of their sequencing, in particular by construction of a cause diagram;

- identification of the faulty or unefficient lines of defense and examination of all related material, organizational and human aspects;

- assessment of risk through examination of potential consequences if other lines of defense fail or if conditions are unfavorable;

- definition of measures to be taken (strengthening or adding of some lines of defense)

- to prevent the occurrence of identical or similar events

- or to limit the consequences of such events

Finally, the analysis report for a significant safety-related event must be sufficiently precise, complete and self-standing so that experience feedback can be used subsequently by other actors.

2.2 Sharing experience feedback and best practice

Central Experience Database and reports on significant safety-related events

To encourage sharing of experience feedback, messages declaring significant safety-related events and corresponding reports are sent to all CEA centers. In addition, the data contained in these documents is recorded in the central experience database (FCE), which can be accessed via an in-house CEA computer network. Processing of this data by the specialists of CEA's Nuclear Safety and Radiological Protection Division (DPSN) results in the publication of annual reports in which general information drawn from analysis of safety-related events identifies roughly ten experience-feedback themes considered as the most important throughout CEA.

Examination of experience feedback from annual reports and conducting internal inspections targeted on these themes allows each facility to find out if it has generic problems and, if necessary, to implement measures necessary to solve them. By consulting the central experience feedback database, each facility can enrich the analyze of its own events through examination of similar events.

CEA/Saclay experience sharing

At CEA/Saclay, sharing of experience relative to incidents is encouraged and promoted during experience-feedback examination meetings, held at least twice a year. These meetings, organized by the Center's safety unit, are chaired by the Deputy Director in charge of nuclear safety. Are also present the CEA/Saclay facility managers and the heads of support departments or other units that play a role in the safety of these facilities.

During this meeting, the facility managers present any event analyses that they have conducted. They are invited to compare their conclusions to lessons learned from experience-feedback reports and to highlight repetitive circumstances. After exchange of views between the various participants, the chairman decides on the various measures to be taken at the Center level.

Other experience-sharing structures

Experience-sharing requires identifying a peer group and creating a structure where peers can periodically exchange views, allowing them to identify common problems, as well as solutions already found elsewhere, and to share incident experience feedback and best practices.

Listed below are some of the particularly active examples of these structures:

- the network of experience-feedback managers,
- the experimental reactor operators' club,
- the criticality engineers' club.

The last two have proven to be so attractive that non-CEA facility operators have asked and obtained to participate.

3. EXAMPLES

3.1 Analysis of two safety-related events

The following section describes two incidents that occurred in the OSIRIS reactor. One concerns an equipment failure and the other is more complex.

3.1.1 Damage to an instrumentation line on the pressurizer in the Isabelle 1 loop

Incident description

This incident, which was declared at the end of 2003 and was below the INES scale, involved an experimental setup for nuclear fuel in the immediate vicinity of the core of the Osiris research reactor. The experiment consists of subjecting fuel rods, placed under the thermodynamic conditions of a power reactor, to neutron flux variations (or "ramps"). (see appendix 4, picture 1)

The experimental setup (Isabelle 1 loop), which is in a bunker set apart from the reactor cavity, consists of a pressurizer designed to regulate loop pressure to 155 bar. (see appendix 4, figure 2)

The incident involved this pressurizer and, in particular, a small-diameter piping system connected to it for passage of an instrumentation line.

During an experiment, automatic water filling was occurring in the loop, indicating a lack of water. Since no other consequences were observed, the experiment was continued up to the end, while increasing the level of surveillance.

After completion of the experiment and after dismantling of the pressurizer, investigations were carried out to determine the cause of the faults.

The incident had no real consequences.

Studies on the potential consequences of incidents of this nature showed that they would remain contained within the reactor building, regardless of any aggravating factors considered.

Detection of the incident and investigations carried out

As indicated above, the incident was detected by monitoring the water level during operation, which proved to be an effective line of defense.

After the experiment, extensive leak tests revealed a through-wall crack, 3 mm long, on the instrumentation line tap on the center part of the pressurizer, as well as other cracks on the associated condensation pots.

An expert metallurgical examination established that these cracks were caused by stress corrosion, initiated outside the piping system. (see appendix 4 – figure 3)

The source of this corrosion was attributed to two causes: on one hand, the high level of stress caused by differential expansion due to a design error in the pressurizer and its instrumentation lines support system (see appendix 4, figure 3), and on the other hand the chemical characteristics of the thermal insulation on the equipment.

Preventive action and experience feedback

Subsequent to analysis of the incident, the facility decided to modify the support system on the pressurizer instrumentation lines to avoid the thermal-stress phenomena resulting from differential expansion on a rigid assembly.

Moreover, the thermal insulation surrounding the pressurizer and instrumentation lines was replaced by a material with controlled amounts of corrosive substances such as chlorine or fluorine.

Periodic inspections of loop integrity were also implemented, and the water level monitoring was clearly identified as safety related.

Other experimental setups of the same type will be inspected, in the context of their next safety review, to ensure that there is no risk of them being damaged in the same way.

3.1.2 Neutron measuring channel positioning error

Incident description

Following prolonged shutdown of the OSIRIS reactor to replace the core vessel, a positioning error occurred during reassembly of the high-level measuring channels on the reactor. This incident was the subject of a report sent to the IAEA (see Appendix 2). As a result, during the first two low-power startup tests on March 19 and 20, 2002, the readings taken to adjust the nuclear power protection thresholds revealed that the temporary value of the "maximum power" threshold, displayed before divergence, was considerably higher than the setpoint value determined at the end of testing.

The protection threshold functions associated with nuclear power were set into operation before the reactor resumed full power, which took place on March 25, 2002.

During the few hours of operation on March 19 and 20, the values recorded in the high-level chambers were approximately 2.3 times lower than the expected values, and consequently, the "maximum power" threshold measured by the neutron chambers in the reactor was temporarily set to 200 MW although, according to technical operating requirements, it should not have exceeded 77 MW. All other protection systems remained operational, however, in particular those that have a more or less direct impact on reactor power, and, since reactor power remained below 40 MW without any rapid power fluctuations, the incorrect setting had no real consequences.

An incident scenario featuring a rapid power fluctuation without an appropriate reaction from the operators could have resulted in cladding failure, leading to an emergency shutdown and contamination of the reactor cavity.

Incident detection and investigations conducted

As indicated above, under-counting in the high-level chambers was detected during the first power escalation on March 19th.

Having attributed this under-count to poisoning of the beryllium located in front of these chambers, it became apparent that the three high-level neutron measuring chambers, that had been removed from their supports during work carried out when the reactor was shut down, had been incorrectly reinstalled with a difference of 5 mm with respect to their original position. Neutron calculations made at a later date showed that this positioning error was consistent with the deviation observed in their calibration.

Preventive actions and experience feedback

As regards the high-level measuring chamber positioning error, the analysis and corrective measures were included in the IAEA report (see Appendix 2, Sections 4 and 5). The preparatory phases for exceptional operations were improved, in particular through coordination meetings between the different actors, formalized by reports, to enable assessment of risks related to operations performed on safety-related equipment..

Subsequently, the occurrence of two incidents declared to the safety authority, in addition to the incident described above, revealed that there was a real need for in-depth analysis of the organization and safety procedures on this basic nuclear installation. This resulted in an audit carried out in September 2002, which led to the publication by CEA/Saclay of recommendations and a pluriannual plan of action, which, in turn, led to a commitment with the Nuclear Safety Authority.

In this plan of action, the essential decisions were to :

- rework the operational organization of INB OSIRIS and, in this re-organization, highlight how "compartmentalization" between the various trades involved could be removed;
- improve information sharing within the facility;
- carry out a relevant, shared and applicable review of documents (in association with the operators of the trades in question) and ensure that its implementation is enforced;
- encourage safety awareness at all levels, along with pertinent questioning attitudes.

This shows that an incident or series of incidents may, in time, require several analyses so that any possible lessons will be learned. The conclusions drawn may vary according to the importance given to one technical aspect or another, or to any given human or organizational factors.

3.2 Generic analysis concerning recurrent events or risk factors

Beyond analyses concerning elementary incidents, the DPSN of CEA or CEA Saclay also carry out more extensive studies based on experience feedback such as typology analysis of recurring incidents or the search for improvement of practices in a given field.

3.2.1 Analysis of incidents involving non-compliance with regulations or procedures

Incidents that occurred in 2003 at the Saclay Center facilities are shown below, classified according to theme, as presented at the meeting with the Nuclear Safety Authority held on April 1, 2004.

	Level 0 (INES)	*Level 1 (INES)*	*Total*
Non-compliance with safety or radiological protection requirements	7	1	8
Insufficient control of operations associated with non-compliance with safety requirements	0	2	2
Insufficient control of operations	1	0	1
Outbreak of fire	1	1	2
Equipment failure	4	0	4
Other	4	1	5
Total	**17**	**5**	**22**

The table shows that for 22 declared incidents, the principal cause of 10 of these was non-compliance with safety or radiological protection requirements. In the other themes, analysis of one of the outbreaks of fire in a unit also revealed acceptance of a pyrophoric component in the unit in violation with entry specifications.

In 11 out of the 22 incidents declared in 2003 (i.e., 50%), one of the causes was non-compliance with a rule of safety or radiological protection.

Based on the study of 17 incidents (the 11 above-mentioned ones in 2003 and six in 2004), CEA Saclay decided to analyze this type of incident more thoroughly, supported by work conducted at CEA level by human-factor specialists from the Nuclear Safety and Radiological Protection Division (DPSN).

The voluntary (V) or involuntary (I) nature of failure to comply with procedures was considered in the study of significant incident reports. An attempt was also made to characterize the type of human action associated with failures: reflex actions (R), predefined actions (P), decision (D). These criteria were then refined by creating an ad hoc classification of recorded failures, which identified seven different types of factors contributing to the observed violations of rules or procedures.

The following table defines this classification by associating the number of cases per category for the above-mentioned 17 incidents covered in this analysis.

Factor of non-compliance with rules	***Nature of action***	***Type of action***	***Number (out of the 17 incidents)***
1 – Rules forgotten (exceptional operation or change of operator, etc.)	I	P	3
2 - Rule not understood by the operator	I	P	5
3 - Rule poorly explained in documents upstream from operator procedures	I	P	3
4 – Rule considered inapplicable (wrong)	V	D	1
5 – Rule considered adaptable (desire to improve)	V	D	6
6 - Error in assessing rule applicability criteria	I	D	2
7 – Error in execution	I	R	4
Total			**24**

Since several factors may be attributed to a single account of non-compliance, the number of factors observed (24) is greater than the number of incidents. (17)

Preventive measures were then considered to prevent the factors defined above. These are:

A. Clarify the decision-making level competent to validate procedures or authorize waivers.
B. Ensure traceability of upstream rules applied in downstream procedures.
C. Involve operators in the preparation or validation of procedures.
D. Involve operators in the preparatory safety and security analyses preceding operations.
E. Teach quality awareness to all personnel (at all organization levels).
F. Teach safety awareness (encouraging a questioning attitude) to all personnel (at all organization levels).
G. Conduct risk analyses (error risks in particular) to establish appropriate checkpoints in procedures.
H. Carry out inspections and audits on compliance with rules and procedures.
I. Conduct technical inspections to test the relevance of procedures.
J. Post instructions at workstations.
K. Improve instruction and training of operators.

The table below shows the correspondence between each measure and the factors that it may prevent.

	1 – Rules forgotten	2 - Rules not understood	3 – Rules poorly explained	4 - Rules considered inapplicable	5 - Rules considered adaptable	6 - Incorrect assessment of rule applicability criteria	7 – Error in execution
A - Clarify the decision-making level competent to authorize waivers				x	x		
B – Ensure traceability of upstream rules applied in downstream procedures.			x	x	x		
C - Involve operators in preparation of procedures or have operators validate them		x		x	x	x	
D - Involve operators in the preparatory safety and security analyses preceding operations.	x	x		x	x	x	
E – Teach quality awareness to personnel	x		x	x	x		
F – Teach safety awareness (encouraging a questioning attitude) to personnel.	x	x	x	x	x	x	x
G – Conduct risk analyses (error risks in particular) to establish checkpoints in procedures.							x
H - Carry out inspections and audits on compliance with rules and procedures.	x		x	x	x		
I – Conduct technical inspections to test the relevance of procedures and practices.	x			x	x	x	x
J – Post instructions at workstations.	x					x	
K – Provide instruction and training for operators.	x	x	x	x	x	x	x

A selection of 4 priority objectives has been debated and validated during the last REX meeting (see Section 2.2 above) with a view to implementing them at the end of 2004 for experimental trials in 2005. These objectives are presented here after each objective is associated with examples of possible actions. These actions are not mandatory: each facility manager has been asked to choose at least one action in front of each objective, with the purpose to progress towards it.

<u>Objective 1: prevent misfits between procedures and practices</u> (good and bad)

- associate the operators to safety and risk analyses
- associate the operator to the writing or to the validation of the procedures
- organize a collective reading of the procedures when preparing the operations
- organize collecting and quick analyses of improves of procedures

<u>Objective 2: improve the pedagogy of feedback experience (REX) at the level of the operators, in each facility</u>

- present and comment REX during meetings associating the operators
- associate the operators to incident analyses
- communicate about REX (including other facilities incidents)
- the area of REX discussion with the operators (safety, radiological protection, classical risks, quality of the facility production and soon)

Objective 3: prevent errors between operating procedures and upstream rules

- quotation of top documents when writing procedures
- identify in the procedures the items resulting from top documents, which the facility manager and personnel are not abilited to change
- organize a step of explicit analysis of upstream documents when creating or modifying a procedure.

Objective 4: preventing the consequences of execution errors by detecting them instead of only try to prevent them.

- analyses possible errors when writing or validating the procedures and organize check steps each time an error is possible and would have a safety impact (loss of a line of defense)
- communicate about human errors and promote neutral checking inside.

Further exchange about these actions will be organized during next REX meetings, and audits of these actions will be performed during year 2005 by the safety unit.

3.2.2 Generic action concerning monitoring of filtering and purification systems

Background

The efficiency of filtering devices (and purification devices, if any) installed on the exhaust from venting systems throughout the nuclear facility is one of the key factors in achieving "radioactive materials containment" safety.

Operators of these facilities are often confronted with problems in this area, as indicated by the number of declared incidents, not including the numerous anomalies for which declarations are not required.

In substance, CEA declared 14 incidents involving filtering and purification systems between the end of 1998 and the beginning of 2004 (nine at Saclay), four of which were given a level 1 classification on the INES scale (two at Saclay).

Starting in 2000, investigations were carried out at certain CEA centers, e.g. at Saclay, following a certain number of recurrent anomalies that occurred in 1999, related, in particular, to inefficiency of HEPA filters and iodine traps in final filtering.

The purpose of this study was to identify the conditions of use and monitoring of this type of equipment and identify the principal causes of the reduced efficiency observed in order to propose improvements.

The nuclear safety and radiological protection division at the CEA level (DPSN)decided to extend this generic study to all CEA centers, of the experience feedback from incidents that occurred between 1999 and 2001.

Work approach

A "Filtering and purification system" work group was set up in the summer of 2003, consisting of approximately ten engineers from different CEA centers, including experts in containment or radiological protection, led by a DPSN "experience feedback" specialist.

The mission of this group was to prepare a feedback experience document to be used by personnel in charge of operating, maintaining and monitoring facilities based on the major anomalies or failures recorded and the "good practices" identified.

The work group, which met four times between September 2003 and March 2004, proceeded on the basis of an in-depth analysis of recorded incidents and studies available in the different centers, completed by a survey carried out with the operators and units in charge of maintaining filtering equipment.

Experience feedback from other major French operators and data from available literature concerning the performance of filters according to the various parameters was also taken into account (US/DOE data in particular).

A first summary document was widely distributed in CEA centers in June 2004.

Experience feedback and recommendations

The above-mentioned document includes 32 recommendations for monitoring filtering and purification systems, grouped according to the different themes useful to facility operators, after presentation of the experience feedback available for each theme.

These recommendations involve operating conditions for filters and iodine traps (storage, assembly, identification, lifetime, etc.) as well as monitoring and efficiency tests to be carried out on them.

DPSN requested strict enforcement by operators of 16 among these recommendations (see Appendix 3 for more information)

The document will be updated and enriched as new data from operating experience or qualification programs and in particular, data concerning the performance of filters in fire situations becomes available.

4. CONCLUSION AND THANKS

Nuclear safety and security require constant vigilance based on a questioning attitude.

Detection of operating incidents is essential to check and, if necessary, reinforce the lines of defense implemented to meet the requirements of defense in depth, without waiting for accidents to occur.

Each manager and each facility learns from its own experience. It is more difficult, but nevertheless essential, to learn from the experience of others as well

I wish to thank the organizers of this conference for giving us the opportunity to do just that.

5. LIST OF APPENDICES

1.- CEA Saclay Center : Research Activities

2.- IAEA Report on the OSIRIS 2002/03/19-20 Incident

3.- Extract from the DPSN document on monitoring filtering and purification systems

4 – Pictures and figures

Organisation and methods used by the CEA Saclay Centre to improve operating procedures and promote best practices in nuclear research facilities

Appendix 1

Presentation of the CEA Saclay centre
Research activities

1. Research in the field of nuclear energy

In this area, research takes place using both simulation techniques and very effective experimental facilities.

Simulation has become so important in research programmes that there are now whole teams working to produce the next generation of computing tools.

Digital simulation of complex systems uses both theoretical and physical models supported by experiment and calculations intended to explore the range of the possible by scientific means, to help in decision-making and to act as an engineering tool for design and optimisation.

This powerful tool, based on developments in computing, is used to produce a precise temporal and three-dimensional description of complex systems and developmental changes. The approaches adopted are multi-scale (from atom to matter, through to the production of useful articles) and multi-physics (mechanical, physical, thermal, hydraulic and physico chemical combined).

In practical terms this research aims to improve the performance of current nuclear power plants, in terms of : Fuel efficiency, particularly during planned changes in the reactor power, optimisation of fuel consumption, life of materials and the operating term of the power plants. It also focuses on the design and selection of future generations of reactors, in terms of long term development.

For long-term stock-piling or storage of radioactive waste in geological strata, simulation techniques can model all the natural events that could possibly contribute to the dispersion of the radioactivity outside the confinement material, through its containers and into the natural environment. This takes into account changes in their physical-chemical properties and transfer mechanisms by diffusion or exchange with underground watercourses.

The Saclay centre has a remarkably varied and high quality choice of experimental facilities available to carry out this research.

Considerable test resources are devoted to mechanical and thermo-mechanical investigations, as well as to thermo-hydraulics of mono-and two-phase (steam vapour) events.

The Tamaris installation, housing the largest vibration table in Europe, is used to test the resistance of materials and large structures to earthquakes. There is a wide range of analytical methods available for the isotopic and physico-chemical characterisation of materials. Corrosion and the degradation of physical and mechanical properties likely to affect materials over time under actual conditions of use, including irradiation, can also be studied. For this, we have at our disposal the OSIRIS research reactor which has been specially designed for irradiating materials, complemented by "the LECI nuclear" laboratory, fitted with remote manipulators, for physical, chemical and metallurgy studies, as well as mechanical tests, on irradiated materials.

These investigations are performed together with CEA's partners, including EDF, Framatome, COGEMA, ANDRA and the IRSN, and for European and international collaborative projects.

2. Technological research

Thanks to considerable progress in the nuclear field, CEA has developed skills that put it at the leading edge of European technical research.

In Saclay, some of the research involves development of instrumentation and non- destructive test procedures, with many applications in nuclear science, aeronautical and automobile engineering, armaments, petrochemicals, new technology in energy or mechanics.

Other research is performed on innovative materials (ceramics, shape memory alloys, nano-materials), and on on-board expert systems, used in networks, transportation or mobile telephones.

The Henri Becquerel National Laboratory (LNHB), a member of the national metrology office, stores and updates the national metrological records on ionising radiation. Its research has, for example, introduced greater accuracy in measuring the radioactive elements present in very small quantities in the environment, or adjustment of radio-therapy dosages.

The CEA also promotes a voluntary technology transfer policy, using partnership research contracts, licensing or help for development of new business companies.

3. Physical sciences

Saclay researchers study matter in a range of different scales and from a variety of angles.

Particle physicists track the smallest constituents of matter, and study the interactions that govern them, in order to develop the unified model of natural forces.

Nuclear physicists try to understand the properties of atomic nuclei, from the simple nucleus to super-heavy and unstable nuclei, that are not naturally present on earth. They are also involved in research on transmutation of highly radioactive waste from the nuclear energy industry.

Astrophysicists are interested in the structure of matter in the universe. How are stars formed, and why do they have planets? Why are stars grouped together in galaxies and how do galaxies evolve?

In these three fields, experimental resources developed in an international context are used : large accelerators for nuclear physics and particle physics (CERN at Geneva, GANIL at Caen, etc.) or space and terrestrial observatories for astrophysics. Saclay is heavily involved in designing and producing these tools. It is internationally renowned for its skills in developing super-conducting coils and magnets and producing ultra-sensitive detection systems.

Other physicists study matter at atomic and molecular level. Over and above their involvement in basic research, understanding these properties will allow new compounds or materials to be designed to meet specific needs in industry or society.

These researches are based on the inter disciplinary application of several basic sciences (atomic and molecular physics, quantum mechanics, statistical physics and the physics of solids, chemistry, biology, etc.) together with a range of instrumentation techniques (large instruments, analysis methods, lasers, accelerators, etc...).

The characteristic shared by all these fields is the handling of microscopic objects on which a range of procedures will be performed to identify their structure or define their reaction to a particular stress (irradiation or corrosion, for example). The information will be collected using physical tools for display and characterisation (electronic microscopes, NMR, micro-probes, neutron beams from the ORPHEE reactor, etc.). Significant developments in modelling are also required, especially in the description of quantum processes and complex phenomena (turbulence, chaos, non-equilibrium thermodynamics, etc.).

The areas of research cover nano-sciences to soft matter, that is, species organised in solution (such as colloids, gels, micelles, etc.), looking at the particular properties of some compounds, such as superconductivity.

4. Life sciences

It is of course essential to study the effects of ionising radiation and toxic materials used in the nuclear industry on living organisms. It involves the detailed investigation of the mechanisms at work at the molecular and cellular level. The data collected will be used to refine the scientific evaluation of the risks of low level doses.

A second line of investigation develops the applications of the technologies stemming from nuclear : tracking and detection of radiation in health and biotechnology and medical imaging that enables the exploration of organs in a non-invasive way. Physicists, doctors and pharmacists work together on the use and development of these techniques in cardiology, neurology, oncology and for studying cognitive function.

Other teams study the structures and functions of proteins in order to create proteins with new properties, particularly useful in medicine. This is protein engineering.

Finally pharmaceutical-immunological specialists develop ultra-sensitive dosage methods. After developing the most sensitive test for detecting mad cow disease they are now finalising a blood test.

All this research involves many scientific disciplines and partnerships with universities and other French research bodies (CNRS, INSERM, INRA, Armed Forces Health Service, public health, Paris hospitals).

Climate and environmental sciences

Saclay researchers are committed to trying to understand the behaviour of climate to predict the changes brought about by human activities.

The models developed to predict situations not previously encountered in the history of the planet are tested on past climate models very different from that of the present day, such as the ice ages.

They can be reconstructed by using isotopic analysis of the ice caps or of marine sediments. Studying recent fluctuations in climate helps in clarifying the mechanisms involved. The part played in environmental cycle of greenhouse effect compounds (carbon dioxide, aerosols, etc.) are also carefully studied. Finally the isotopic tool is used to date past events and study transfer of matter in today's environment.

Appendix 2

IAEA Report on the OSIRIS 2002/03/19-20 incident : Positioning error for the reactor neutron detectors

IAEA Number: FR-0014/2002_03_19/01	Date of receipt: November 2003
Title: Positioning error for the reactor neutron detectors	
Country: France	Date of incident: 2002/03/19-20 Type of report: Main Follow-up expected: No
Name of research reactor: OSIRIS	Power: 70 MW
Research Reactor Code: FR-0014	Designer: CEA
Research Reactor TYPE: Pool-type	Start of operation: 1966

Abstract

After a long shutdown of the reactor OSIRIS, a positioning error occurred during the reassembly of the "high level" neutron detectors. This resulted in an operation of the reactor, on the 19th and 20th of March 2002, during a cumulated time period equivalent to 13h30, with "maximum power" safety thresholds adjusted to a 200-MW power, instead of the 77 MW-power authorized by the technical specifications. These thresholds are one of the initiators in 2/3 of the reactor emergency shutdown. It should be noted that other lines of defense as regards potential risks linked to this incorrect adjustment of the safety neutron flux thresholds remained active.

This incident underlines the need for a good preparation and an exhaustive identification of the risks associated with rare or infrequent operations possibly affecting the reactor protection system. In this case, procedures are often incomplete or inexistent. Stress has to be put on requalification, which has to be adapted to the operation.

2- Narrative description (see figure 1)

OSIRIS reactor is a pool-type reactor. The core of this reactor comprises 38 fuel elements made of U_3Si_2 plates enriched with 19.75% ^{235}U and 6 control elements. Its authorized rated thermal power is 70 MW.

The power of OSIRIS reactor is monitored:

from the subcritical operating conditions to a power of about 1 MW using three neutron measuring chains equipped with "low level" fission chambers.

beyond 1 MW using three neutron measuring chains equipped with "high level" compensated ion chambers.

Neutron measuring chains may be used to measure the reactor power only if they are calibrated. To do so, they are routinely checked and compared to the power deduced from the thermal balance and the power determined by the nitrogen -16 measurement chain.

These neutron measuring chains are connected to the reactor protection system. There are also two other means used to measure the reactor power, either using the measurement provided by the regulation chain equipped with an ion chamber or using the nitrogen-16 activity measurement of the core cooling water. Lastly, the reactor power is calculated using a thermal balance relating to the primary cooling system; this measurement requires a stabilization period of one hour approximately. Its precision is good over 40 MW.

During summer 2001, OSIRIS reactor was shutdown for a period of 8 months approximately, in order to replace the core tank. In this work, most of the equipment installed in the reactor pool, and especially the "low level" and "high level" neutron detectors were disassembled. During this exceptional operation, the neutron detectors were physically removed from their specific position but remained connected.

During the reassembly of the "high level" neutron detectors on December 3rd, 2001, the operators (different from those who disassembled the neutron detectors) realized that there were several possible positions to reassemble them. Indeed, the lead protection of the concerned detectors is secured with two sets of pads fitting together in two holes of the fixing support which has several holes each separated by 5 cm. Observing this table, they noticed the existence of two 5-cm apart marks corresponding to two positions that the neutron detectors had in the past. As the mark the most distant from the reactor core was the most visible, the operators deduced that it should be the most recent and then trustfully placed the lead protection of the neutron detectors on these marks.

Neutron detectors were then reassembled and their thresholds were kept to the last values introduced before the work. The "high level" neutron chains were tested using these thresholds.

The first reactor start-up was followed by a low power operation of the reactor on the 19th of March 2002 in the morning. During the low power operation, the reactor protection system was connected to the "low level" neutron chains.

The second reactor start-up, on the 19th of March 2002 in the afternoon, was followed by a 31 MW power reactor operation. During the power rise, around 1 MW approximately, the operating team noticed that electrical signals at the output of the "high level" chains were lower than the values measured before the work. The low signal from the "high level" chains was then attributed to a poisoning of the beryllium behind which the neutron detectors are located. The operating team considered that the reactor power remained controlled by other measurements (nitrogen-16 chain and thermal balance). The reactor power increase was pursued in stages.

A third reactor start-up, was followed by a 40-MW power reactor operation, on the 20th of March 2002, under similar conditions.

On the 25th of March 2002, the operator modified the reactor emergency shutdown thresholds associated with the maximum power of the "high level" chains using results of the thermal balance performed on the 20th of March 2002, as a basis and the reactor was restarted for a new operation cycle.

However, the investigations conducted to determine the cause of the low signal from "high level" neutron detectors showed a 5-cm error on the positioning of these detectors with regard to the reactor core. The calculations performed showed that this positioning error resulted in a reduction of the magnitude of the electrical current from these detectors by a 2.3 factor, as observed.

Investigation of the unusual event and safety assessment

The reduction by a 2.3 factor of magnitude of the current of the neutron chains resulted in an adjustment of the "maximum power" safety thresholds to a 200-MW value instead of the 77-MW value fixed by the technical specifications. The inappropriate adjustment of protection thresholds associated with the "high level" neutron chains distorted the reactor protection upon excessive maximum power during two operation periods of the reactor over a total time period of 13 hours and 30 minutes approximately (5 hours and 30 minutes on the 19th of March 2003, and 8 hours on the 20th of March 2002). However, all the other reactor protections remained operational, especially protection against inadvertent power excursions (through the power doubling time thresholds) and against high power levels (through the water temperature thresholds at the core outlet). Besides, power changes were monitored during this period by other systems such as the nitrogen-16 chain and the thermal balance performed for the primary cooling system.

It should be noted that during this time period, the real reactor power remained below 38 MW, and as there was no reactor power disturbance, this incident did not have any real consequence as regards safety.

Protection thresholds were adjusted to their specified values on the 25th of March 2002.

In case of an inadvertent power excursion, the other lines of defense as regards potential risks linked to this incorrect adjustment of the neutron flux safety thresholds are:

Trigger of the minimum threshold (adjusted to 3 seconds) associated with the neutron flow doubling time, resulting in an automatic shutdown of the reactor.

Automatic shut down of the reactor by the cladding failure detection system in case of a too high temperature reached in the fuel clad.

Automatic shut down of the reactor if the maximum temperature threshold at the core outlet is reached due to insufficient cooling of certain fuel plates at high reactor power.

Observed causes and corrective actions

As the core tank was replaced, the three « high level » neutron detectors were disassembled and reassembled in a position different from expected. This incorrect positioning is mainly due to causes indicated below.

Neutron detectors disassembly and reassembly operations, which are unusual operations, were insufficiently prepared. The procedure related to these operations was not precise enough.

Communication between the operators responsible for writing the maintenance procedures and those responsible for applying them was not sufficient.

A common cause may result in an incorrect positioning of the three neutron detectors, the single lead protection.

There was no requalification of the "high level" neutron measuring chain before the reactor restarted.

The operating team early noticed the difference between electrical currents from the "high level" neutron detectors, but it did not adjust directly the "high level" maximum power thresholds using the thermal balance. According to the operator, this resulted from the lack of precision of the thermal balance below 40 MW and from the beryllium poisoning assumption. According to the operator, the operating team voluntarily reduced the reactor power in order to better understand the phenomenon.

The immediate corrective action was the adjustment of the above-mentioned thresholds on the 25th of March 2002. Since then, the reactor operates in compliance with its safety documents.

Lessons learned

This incident highlighted the need to improve the preparatory phases of exceptional or infrequent actions performed in the facility. A coordination meeting between the different operating staff, formalized by a report, shall be scheduled to study the risks arising from the planned operations, precautions to be taken in order to minimize these risks and the requalification of the concerned equipment. Moreover, the operating staff will be made aware of the risks arising from the actions performed on equipment important for safety and the care which must be taken for their requalification.

In order to prevent a repetition of such an incident, the disassembly and reassembly procedure for "high level" neutron detectors has been completed so as to specify the exact location of these detectors.

Coded watchlist

1 -	Reporting category:	1.2.4 1.3.3
2 -	Status of the reactor prior to the event:	2.5
3 -	Failed/affected systems:	3.4.1 3.4.2 3.4.3
4 -	Failed/affected components:	4.1.9
5 -	Cause of the event:	5.1.1.9 5.1.10.2 5.3.1 5.4.5. 5.5.7 5.5.9.2
6 -	Effects on operation:	6.0
7 -	Characteristics of the incident:	7.4 7.6
8 -	Nature of failure or error:	8.3
9 -	Nature of recovery actions	9.1

Figure 1

Relative position of the OSIRIS reactor core
and associated neutron detectors (high power level)

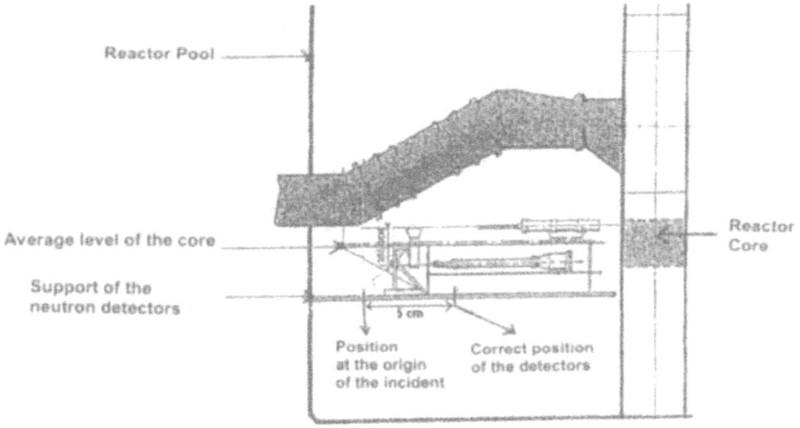

ORGANISATION AND METHODS USED BY THE CEA SACLAY CENTRE TO IMPROVE OPERATING PROCEDURES AND PROMOTE BEST PRACTICES IN NUCLEAR RESEARCH FACILITIES

Appendix 3

Recommendations based on operating feedback concerning acceptance, installation, monitoring and maintenance of filtration and purification systems

Selected recommendations to be applied on a mandatory basis (List taken from document DPSN/SSN/FT/022 - Rev. 0 - 18 May 2004)

Recommendation no. 5

FILTER INSTALLATION AND VERIFICATION PROCEDURE: Each facility must absolutely make sure that filter or iodine trap replacement operations and post installation checks are performed in accordance with validated and approved procedures documented in the general operating rules. Such procedures (and specific hold points) are intended to prevent installation errors or use of incorrect filter or iodine trap types (verification of filter medium or active carbon compliance with safety baseline requirements by operator). The type of equipment to be installed (parts list) must therefore be specified in the filter and iodine trap replacement procedures.

In the case of equipment modified for use in aggressive atmospheres (for example, change of filter seal type, frame or support material, etc.), specific marking or labeling should be used to inform maintenance teams and prevent incorrect filter type errors during installation.

Recommendation no. 9

LIFETIME OF NON-FIRST BARRIER VHE FILTERS: Given the operating feedback and bibliographic data currently available, it is recommended that the lifetime of VHE filters (including those approved by CTHEN) be limited to a maximum of 15 years for all CEA facilities, old and

new, with the exception of first barrier filters installed in gloveboxes, shielded containments and process ventilation systems. This situation could be achieved in two or three years for existing facilities on the basis of specific action plans for each center. In the case of facilities nearing decommissioning (less than 5 years) or currently undergoing remediation or dismantling, a longer lifetime could be tolerated on the basis of a specific risk analysis.

This lifetime could be subsequently adjusted, if necessary, based on available operating feedback (including detailed visual inspection of oldest replaced filters, and specific behavior tests to be defined).

Recommendation no. 10

ADEQUACY OF VENTILATION SYSTEM OPERATING CONDITIONS WITH RESPECT TO FILTER SPECIFICATIONS: Adequate measures must be adopted in facilities so as to ensure that the real operating conditions of ventilation systems are compatible with filter media specifications (in terms of relative humidity, temperature, differential pressure, etc.). In particular, experimenters must verify and control the impact of processes implemented inside the containments, as they can affect the nominal characteristics of the air to be filtered.

Recommendation no. 11

PERIODICITY OF GLOVEBOX EXTRACTION FILTER REPLACEMENT OPERATIONS: The following general principle shall be adopted concerning the definition of replacement frequencies for glovebox/cell extraction filters (ventilation type IV or IIIB, containment class C4, areas with high or extreme radiological risks) and process ventilation first barrier VHE filters:

- *Shorter periodicity than that specified in ISO 11933-4 standard (2 years maximum for extraction filters) if imposed by operating feedback, conditions of use and specific risks associated with the process implemented in the glovebox or containment.*

- *Longer periodicity than in said standard provided that it is possible to either periodically test the efficiency of the filters (testing after each filter replacement operation, as opposed to yearly systematic testing with possibly limiting effects due to large number of glovebox filters contained in a facility) or monitor their efficiency using reliable indicators such as the atmospheric contamination level downstream of the first barrier; in any case, the maximum lifetime of glovebox extraction or process ventilation first barrier filters (most exposed to aggressive atmospheres) must not exceed 5 years.*

- *Various solutions can be considered: Equipping the ventilation pipes at injection and sampling points with an air sampling system and offline counting system so as to test the first barriers after each filter replacement or installation operation upstream of the first barrier.*

- *If the implementation of such indicators is not possible, a periodicity in accordance with the standard should be adopted.*

Recommendation no. 14

FILTER REPLACEMENT TIME IN CASE OF CLOGGING: In the event of load losses exceeding the limits specified in the operating instructions (negative pressure limits to be maintained in the various areas), the filter must be changed within one month of fault detection. This delay can be shortened for certain facilities, depending on the potential risks involved (particularly in case of rapid and predictable development of clogging due to process, periodic ventilation transients, etc.). This delay must be specified in the periodic inspection procedure (specific instructions). In case of absence of an auxiliary filter (bypass connection), compensatory measures shall be implemented if necessary during the interval between filter inspection and effective replacement.

Recommendation no. 15

AIR FLOW MEASUREMENT CONCURRENTLY WITH LOAD LOSS MEASUREMENT: For facilities where the air flow in ventilation pipes is likely to vary (progressive drifts) with respect to nominal conditions, the load loss value measured at the filter boundaries must be accompanied by a measurement of the air flow effectively passing through the filter at that moment. Minimum basis: yearly airflow measurement at DNFs and, if possible, at the intermediate filter barriers.

Recommendation no. 18

LOCATION OF INJECTION AND SAMPLING POINTS: For each system to be tested, the location of the injection and sampling points should be verified so as to ensure an adequate representativity and homogeneity of the mixture for each sample. Their proper positioning must be plotted. Regarding the adequate mixing distances recommended in standards NF X 44-011 and NF M 62-206:

If the filter unit design allows it, a minimum distance of ten times the pipe diameter shall be systematically observed between the injection point and the sampling point upstream of the filter or trap.

A mixing distance comprised between 5 and 10 pipe diameters shall be considered as acceptable provided that the various measurements involved are performed under strictly reproducible conditions (year after year), particularly as regards the positioning of sampling probes in ventilation pipes.

If the mixing distance is less than five pipe diameters, the use of a multiple jet injector method shall be privileged. If necessary, additional tests shall be performed to ensure the correct homogeneity of the gas flow at the sampling point upstream of the filter (for example, increasing the measuring points in the sampling section, or plotting grid lines (rectangular section) or quarter-crown divisions (circular section) and sampling at the center of the regions thus determined).

Recommendation no. 19

VALIDATION OF INJECTION POINTS: Helium tracing homogeneity measurements shall be systematically performed after each significant modification of filter systems or in case of doubts concerning the representativity of samples taken from existing filter systems (particularly in the case of old facilities). If necessary, these measurements shall be used to determine the minimum mixing length (based on system characteristics) below which it shall be necessary to use multiple jet injectors. In case of clearly insufficient homogeneity, a specific study must be performed, possibly leading to system modifications or injection and sampling point repositioning.

Recommendation no. 21

MEASUREMENTS DURING TESTING: The test procedures must be strictly observed so as to minimize the degree of uncertainty associated with the facility itself and possibly affecting the filter and iodine trap real efficiency values, and to avoid erroneous points. In particular, to ensure compliance with standards NF X 44-011 and NF M 62-206, relative humidity (average value during test) and ambient temperature must be measured systematically at upstream and downstream injection and sampling points. These measurements shall be used to verify that the thermohygrometric equilibrium of the filtration or purification system has been attained and that each measurement has been performed under the same conditions as previous measurements (to ensure that the efficiency values obtained are applicable and comparable with previous values). The air flow passing through the filter or trap must also be measured. The values read shall be used to interpret possible variations and to determine associated uncertainties. These values and the global uncertainties associated with each measurement must be included in the test report.

Relative humidity measurements shall be performed in accordance with standard NF X 15-010. Ventilation flow measurements shall be performed by assessing the flow rate in a cross-section of the pipe (Pitot or anemometric probe) in accordance with standards ISO 3966 et NF X 10-112, with a maximum relative error of 10%, or using a tracing method if the other methods cannot be applied.

Recommendation no. 23

CHOICE OF TRACER FOR IODINE TRAP TESTS: In order to minimize discharges, only one iodine trap efficiency test shall be performed per facility. The choice of tracer type (I_2 or ICH_3) must be based on the operating conditions of the facility and, if applicable, on the physicochemical form most easily discharged under accident conditions. Nevertheless, a molecular iodine test may be recommended for inspection purposes after an iodine trap replacement operation (to check proper installation).

Recommendation no. 26

ACKNOWLEDGEMENT OF POTENTIAL RISKS OF RADIOACTIVE IODINE DISCHARGE DURING IODINE TRAP EFFICIENCY TEST: In order to anticipate changes in regulations concerning gas discharges from CEA facilities, all basic nuclear installations must systematically address a request for prior approval to the SPR/Environmental surveillance division before each iodine trap efficiency test. This agreement is subject to favorable meteorological conditions (optimal diffusion) minimizing the potential radiological impact in case of accidental discharge of iodine 131 due to insufficient iodine trap efficiency. Sufficient margins must be provided to allow for possible postponement of tests (in case of unfavorable meteorological conditions) with respect to imposed deadlines based on facility-specific instructions. It is also recommended that potential discharges inherent to iodine trap efficiency tests be properly acknowledged in discharge and water sampling orders for all basic nuclear installations.

Recommendation no. 28

SCHEDULING OF FILTER REPLACEMENT OPERATION (FURTHER TO DETECTION OF INSUFFICIENT EFFICIENCY) AND SUBSEQUENT REPLACED FILTER EFFICIENCY TEST: In case of non-observance of the minimum efficiency value imposed for a non-redundant filter or iodine trap, the equipment must be immediately replaced and its efficiency must be tested immediately after replacement (see RFS RR1). A filter bank subject to a filter replacement operation cannot be used again until it has been requalified. It is therefore necessary to have a minimum stock of filters and iodine traps available and to schedule the replacement operation and corresponding efficiency test on the same day. If immediate replacement is not possible (due to physical constraints) or the filter bank has not yet been requalified and considered as newly available, the installation or unit shall be set in shutdown position and compensatory measures shall be implemented. If necessary, the ventilation system shall remain shut down and all operations capable of contaminating the atmosphere in corresponding containments and rooms shall be forbidden (particularly experiments susceptible of producing radioactive iodine).

In terms of design, the implementation of an additional filter bank in the ventilation system allowing for normal operation when a bank is unavailable should be systematically studied for all installations currently being designed or upgraded.

In the case of a redundant filter or iodine trap, non-observance of the minimum imposed efficiency value must lead to its immediate transfer to the auxiliary filter bank. The filter or iodine trap replacement operation in this bank should be performed within a maximum of 72 hours (subsequent efficiency test within 24 hours). The requalification delay must be minimized to allow for potential degraded or accident conditions requiring the use of this auxiliary bank. In the case of iodine traps, if the quantity of iodine 131 immediately available is not sufficient and the center does not have the means to prepare and fraction the sources required (determined according to characteristics of system to be tested), the procurement time for the radioactive iodine required for the tests may exceed 72 hours and therefore generate a constraint (for example, the minimum procurement time for a 1 MBq source is approximately two weeks for the PHENIX facility). In this case, the organization implemented should allow for a minimal delay between the iodine trap replacement operation and the subsequent efficiency test.

In the case where the measured efficiency for a non-redundant filter or iodine trap is between 1000 and 2000, corresponding to the preventive efficiency low threshold currently adopted (see section 3.6.5.1 below), the replacement operation shall be performed as quickly as possible, within a maximum of one month (if the change in efficiency level since the last test is considered sufficiently slow). The efficiency test shall be performed immediately after the replacement operation. In the case of a redundant filter or iodine trap, its transfer to the auxiliary bank shall be performed immediately and the replacement operation shall be performed within one month (and subsequent efficiency test within one week).

These recommendations are summarized in the table below so as to correctly distinguish all possible cases.

	Filter (or iodine trap) efficiency during test	*Filter (or iodine trap) replacement time and delay between replacement operation and efficiency test*
Non-redundant filter (or iodine *trap)*	*E < 1000*	*Immediate replacement (in case of physical impossibility, implementation of compensatory measures such as interruption of operations with risk of contamination dispersion)* *Execution of efficiency test immediately after replacement*
	1000 < E < 2000	*Replacement as quickly as possible and within a maximum of one month (in case of slow changes in efficiency level since last test)* *Execution of efficiency test immediately after replacement*
Redundant filter (or iodine trap)	*E < 1000*	*Transfer to auxiliary bank* *Replacement within 72 hours and efficiency test within 24 hours*
	1000 < E < 2000	*Transfer to auxiliary bank* *Replacement within 1 month and efficiency test within 1 week*

Recommendation no. 29

ANALYSIS OF RISKS OF AGGRESSION: It is absolutely necessary, during operation (existing facility) or during the design phase (new filtration or purification unit), that operators, maintenance personnel and designers systematically perform an analysis on the basis of information supplied by experimenters (changes in processes implemented, etc.) so as to:

- *Identify parameters possibly affecting filter and iodine trap efficiency, i.e.: factors favoring slow or rapid changes in efficiency level, operating conditions potentially generating aggressions for these systems, etc.*

- *Implement optimal preventive actions for the parameters identified*

Recommendation no. 30

MAINTENANCE POLICY ORGANIZATION AND IMPLEMENTATION : It is absolutely necessary to implement a preventive maintenance policy for purification systems based on a systematic analysis of the results of efficiency tests performed on DNFs so as to identify all abnormal behavior or generic faults in filter elements, optimize their inspection frequency, anticipate their replacement on the basis of clearly defined criteria (see section 3.6.5.1 of document DPSN/SSN/FT/022 - Rev.0 - 18 May 2004) and adopt necessary corrective measures to ensure a system efficiency value at least equal to the minimum value required until the next periodic test.

A corresponding organization shall be implemented for each installation to ensure proper monitoring, operation and traceability within the efficiency test intervals:

Maintaining measured efficiency values, test parameters and ventilation operating conditions for an appropriate period at least equal to the filter or iodine trap lifetime

Recording all causes identified possibly explaining observed drops in efficiency leading to filter or iodine trap replacement (for subsequent analysis)

Recording all difficulties encountered during execution of tests (operating feedback)

Recommendation no. 31

IDENTIFICATION OF DNFs FURTHER TO MODIFICATIONS OR NEW WORK: It is necessary to verify (based on an exhaustive analysis, particularly further to new work or modifications) that all filter barriers and iodine traps installed in the various ventilation systems (process, gloveboxes/cells, building, effluent tanks, specific air injection systems, etc.) fulfilling a DNF function (in normal operation, incident or accident situations) and therefore inspectable on a yearly basis have been properly identified as such in the safety baseline.

<u>*Recommendation no. 32*</u>

PROCESSING OF NON-OBSERVANCE OF MINIMUM EFFICIENCY CRITERION: All identified cases of non-observance of the minimum efficiency value imposed for DNFs must be systematically recorded in a fault sheet and the corresponding causes must be searched for.

Assuming that each installation has a clearly established maintenance policy and an organization capable of detecting abnormal filter or iodine trap behavior and anticipating their replacement on the basis of clearly defined criteria before the equipment has lost its efficiency, this type of fault <u>shall be systematically covered in a subsequent report to the Safety Authority</u> within the scope of the yearly safety reports for basic nuclear installations (it must be noted that the number of faults of this type should normally decrease in years to come).

These faults shall only be declared as significant incidents to the Safety Authority (in accordance with applicable procedures, see NIG 472) if one or more technical, human or organizational errors are identified possibly corresponding to the non-observance of another DNF safety-related requirement, in addition to the observed efficiency fault, or possibly explaining said efficiency fault (for example, installation of HE filter instead of VHE filter, non-observance of yearly inspection periodicity, etc.).

Appendix 4

Pictures and figures

Picture 1 – Osiris research reactor is used to perform qualification of new nuclear fuels and behavioural studies for structural materials exposed to neutron radiations, and qualification of computer simulation in these areas.

Figure 1 – Isabelle 1 loop is associated to Osiris research reactor and is focused on fuel tests needing rapid variations of neutron flux. A safety, related incident occurred on the "pressuriseur", located in the "casemate". A loss of coolant was detected in the "cuve à niveau".

Figure 2 – After the experiment, extensive leak tests revealed a through wall crack, 3 mm long, on the instrumentation line tap on the center post of the pressurizer, as well as other cracks on the associated "pots de condensation".

Figure 3 - An expert metallurgical examination established that these cracks were caused by stress corrosion, initiated outside the piping system. The source of this corrosion was attributed to two causes: on one hand, the high level of stress caused by differential expansion due to a design error in the pressurizer and its instrumentation lines support system and on the other hand the chemical characteristics of the thermal insulation on the equipment.

Picture 1 : Osiris research reactor

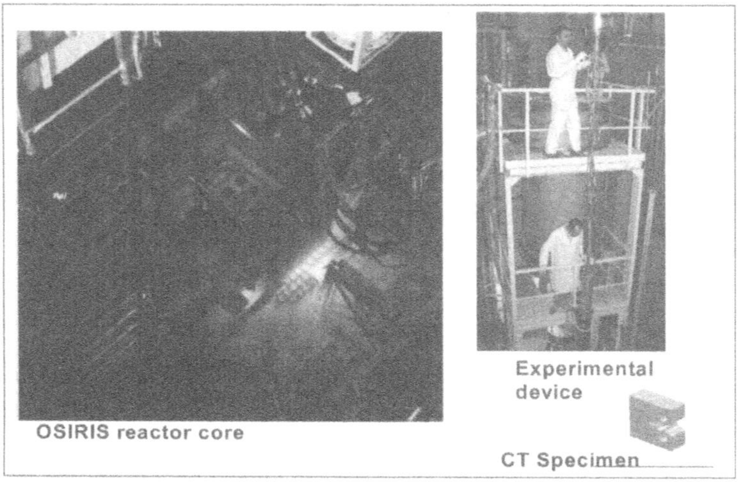

Figure 1 : Isabelle 1 loop (experimental setup)

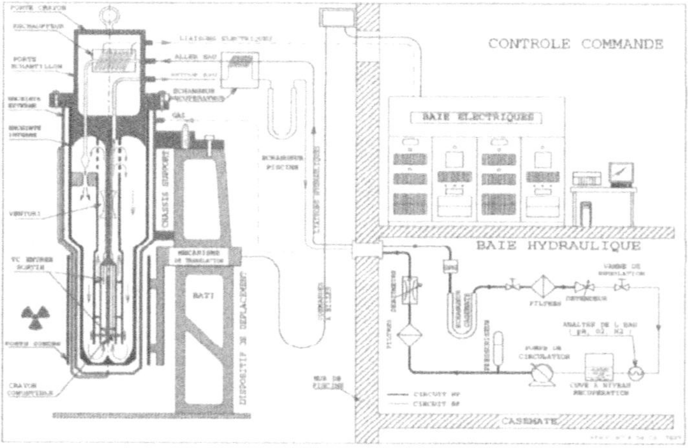

Figure 2 : diagram of pressurizer with leakage localization

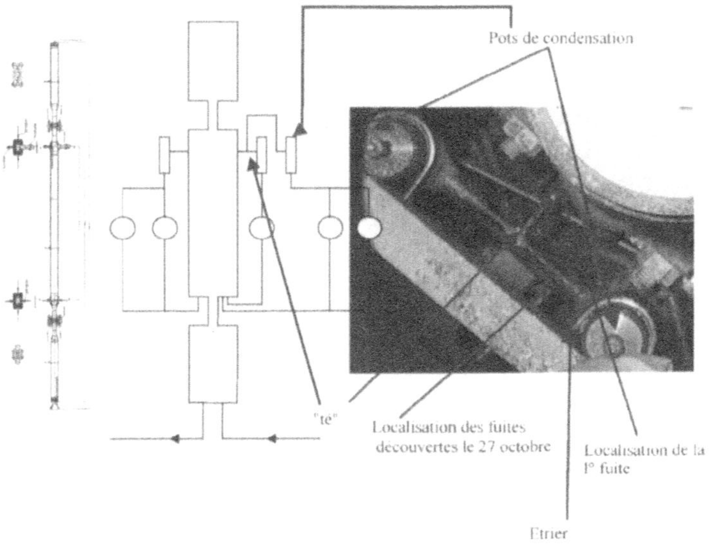

Figure 3 : cracks caused by stress corrosion

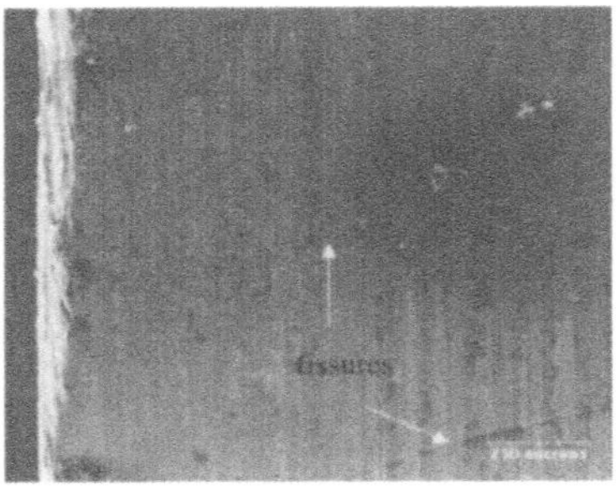

Fissures sur la surface externe du tube

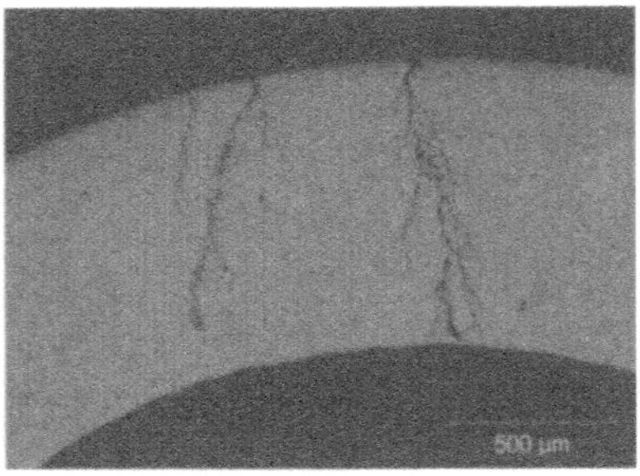

Fissures caractéristiques de corrosion sous tension

SAFETY AS AN UNCEASING PROCESS: THE ROLE OF THE HEALTH PHYSICS & SAFETY DEPARTMENT IN A NUCLEAR RESEARCH CENTRE

Pascal Deboodt
SCK-CEN, MOL

Abstract: The nuclear research centres present a lot of characteristics which require from the people in charge of the Health Physics and Safety Department an appropriate approach. After the description of these characteristics and based on fifteen years of practice, using some real events leading to incident or quasi-accident, our contribution will indicate the approach which has been implemented, the main results gained and the trends for the next years. Although mainly focused on radiological risks, discussion will also be provided concerning the need for a global management of risks on the workplace, this topic being more and more considered as a fundamental for insuring the safety of the workers.

Key words: Optimisation – Risk Management.

1. INTRODUCTION

1.1 The SCK•CEN: missions – installations

With a view to sustaining development by research and development, training, communication and services, SCK•CEN contributes to:

- Nuclear safety and radiation protection;
- Medical and industrial applications of radiation;

F. Lambert and Y. Volkov (eds.), Safety Improvements through Lessons Learned from Operational Experience in Nuclear Research Facilities, 187–196.

- The backend of the nuclear fuel cycle.

Its mission implies the extension of conventional activities in two important fields.

- The non-energetic applications of nuclear energy are becoming increasingly more relevant to society, especially in the medical sector;
- Sustained development implies that non-technical aspects, such as social and economical factors, are also taken into account.

In order to perform its mission, the SCK•CEN has the advantage of the existence of the following departments:

- Reactor and fuel safety;
- Radioactive waste;
- Radiation protection;
- Training;
- New and societal topics.

These departments perform their tasks by means of a large diversity of installations such as BR1 (4 MWth graphite-moderated, air-cooled reactor), BR2 (high neutron flux reactor), VENUS (zero power critical facility), BR3 (prototype of the PWR's; now under decommissioning), HADES (underground laboratory at a depth of 225 m for the study of clay as potential geological host formation), LHMA (effects of irradiation on materials) and other nuclear analysis and chemical laboratories.

1.2 Characteristics of the Belgian Nuclear Research Centre (SCK•CEN)

- As presented here above, there is a quite large diversity as far as the installations and the mission of the SCK•CEN are concerned. Moreover, the actual situation of these installations shows a large spectrum of operational conditions.
- These technical characteristics must be completed by some other such as:
  - The large variety of people employed at the SCK•CEN: workers, staff members, students, interns, visitors, external workers;
  - The role played by the centre as support of the Regulatory Body (the Belgian Federal Agency for Nuclear Control), as well as its contribution to the public information.
- Last but not least, during the last decades the SCK•CEN faced the implementation of new regulations. These regulations are dealing with

radiation protection at the workplace (Royal Decree of July 20, 2001) and welfare of the workers at the workplace (Royal Decree of August 4, 1996). These regulations are the translation of international recommendations (such as the ICRP-60, 61 Publications and the European Directives EUR/96-29 and EUR/97-43 concerning the radiological risks, or the European Decision DIR 89/391/EEC related to health and safety at the workplace).

2. THE HEALTH PHYSICS AND SAFETY DEPARTMENT

2.1 General structure

As a result of the regulations, the actual structure of the Health Physics and Safety Department at the SCK•CEN looks like the simplified presentation given by figure 1.

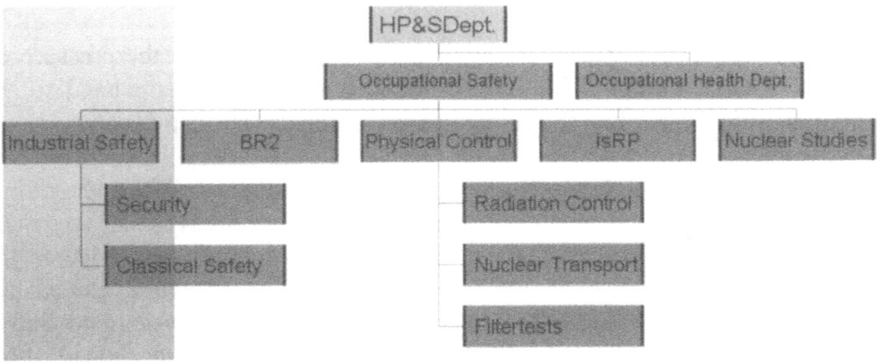

Figure 1: The Health Physics and Safety Department of the SCK•CEN

The head of this department reports directly to the general manager and for his nuclear responsibilities he's controlled by an independent control body which reports to the Belgian **F**ederal **A**gency for **N**uclear **C**ontrol.

This structure shows the integration of the occupational and health aspects as well as the industrial safety aspects. It also clearly appears that the service "Physical Control", which is required by the Royal Decree of July

20, 2001, is in charge of the radiological part of the safety and security at the SCK•CEN.

2.2 Available tools

2.2.1 Legal tools

Beside the Health Physics and Safety Department described here above, there are also other legal tools at the SCK•CEN:

- The **I**nternal **C**ommittee for **P**revention and **P**rotection at the **W**orkplace; this committee is constituted by representatives of the hierarchy, trade unions and of the occupational health department. The head of the Health Physics and Safety Department is the secretary of the ICPPW.
- Inspections performed by independent control bodies; there are two main bodies acting for the control at the SCK•CEN. One is in charge of the nuclear installations; another is dealing with conventional installations (lifts, ladder, electrical circuit,…).

2.2.2 Other tools

Some other tools have been implemented, as a result of the pressure on the field (see section 3) or as a consequence of decisions related to the internal policy regarding the safety at the workplace. These tools are:

- The ALARA and Safety Committee; this committee, reporting directly to the general manager, is constituted by representatives of the main departments of the SCK•CEN. Its mission is to provide advice to the operators on basis of the technical and ALARA procedures established by him and introduced for approval by the Health Physics and Safety Department. It's worthwhile to note that also, at this level, both radiological and non radiological risks are taken into account.
- The Advisory Committee for Safety; this committee is composed by representatives of the trade unions and is mainly involved in case of operations presenting new risks or performed for the first time.
- The international school for Radiological Protection (isRP); its mission is to provide adequate training to all the people that are employed at the SCK•CEN. The school is also increasingly active for the training of other categories of people as support for the FANC or the European Commission for example.

- Safety visits of all the workplaces at the SCK•CEN; these visits occur twice a year. The first one is meant to check the situation, to detect the unsafe conditions (devices)... The second one has to evaluate the evolution and the improvements required by the report of the first visit. Both visits lead to reports which are distributed to the members of the ICPPW.

Moreover, some specific tools have been developed in order to take into account the characteristics of the centre as described here above. For example:

- Mock-ups of installations;
- Software VISIPLAN (see figure 2);
- ALARA procedure.

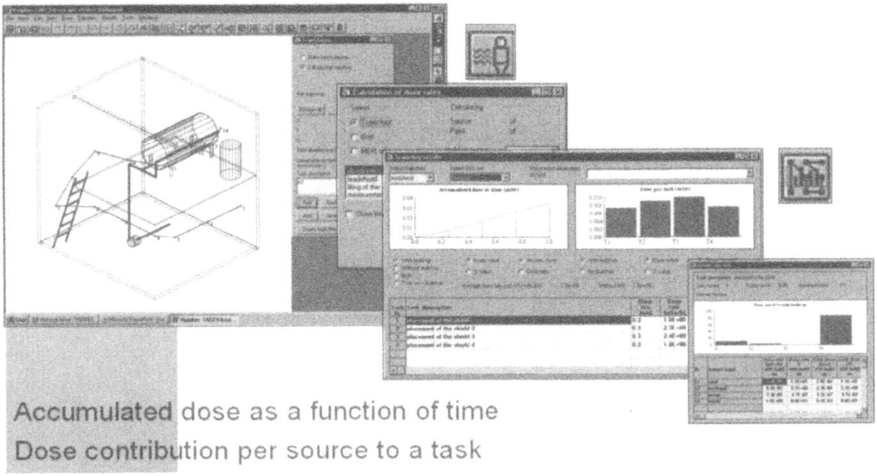

Figure 2: VISIPLAN

As it can be concluded from the previous lines, the Health Physics and Safety Department faced some evolutions of the legal frameworks as well as of the operations performed at the SCK•CEN installations.

2.3 Some results

Figures 3 to 6 provide the main relevant results of the approach which has been implemented at the SCK•CEN. It's worthwhile to make the following comments:

- Since implementation of the ALARA procedure, the collective doses as well as the maximum individual dose are significantly reduced. If this conclusion seems to be expected, one has to keep in mind the characteristics of the installations and the consequences ofsuch yard as a decommissioning project of a PWR.
- Quite surprising, there is also a decrease of the number of the "conventional" accidents. For classification, such accident is defined as an event leading to at least "one day off" for the worker due to his injuries.
- Having a look at the data for BR3, we can find a good example of the implementation of the ALARA principle for a long period of time. Indeed, if higher doses were received during the decontamination of the primary loop (figure 4), the averted collective dose during the following years was estimated as high as 7 man.Sievert!
- The level of maximum individual dose at BR3 has reached a value which lies in the range of the Belgian natural background!

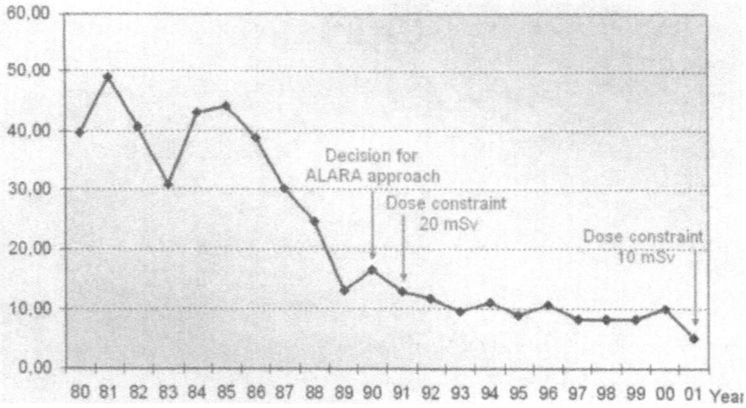

Figure 3: Maximum individual dose SCK•CEN (mSv)

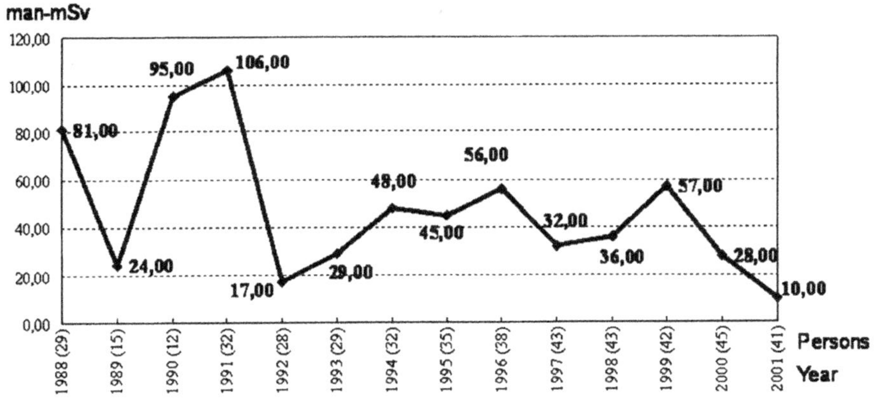

Figure 4: Annual collective dose BR3 (man.mSv)

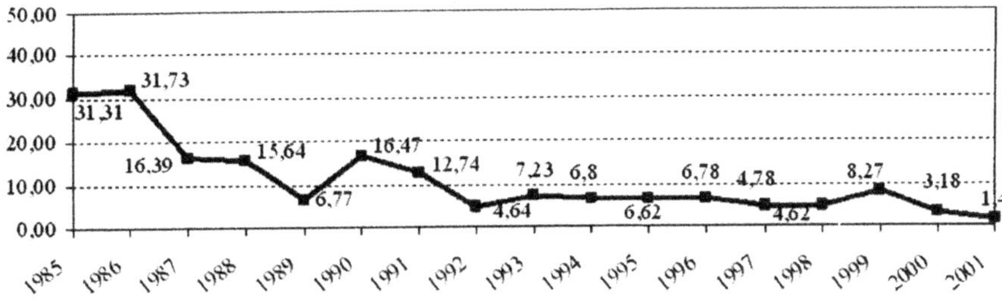

Figure 5: Maximum individual dose at BR3 (mSv)

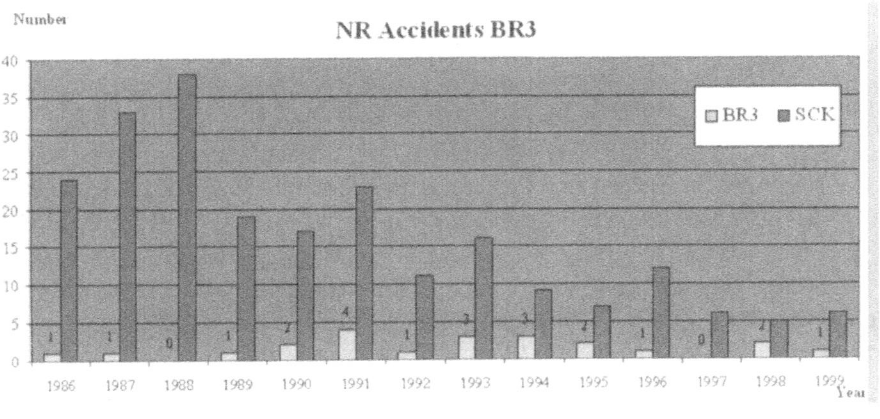

Figure 6: Number of non radiological accidents at BR3 1986 - 1999

3. THE PRESSURE OF THE FIELD

As mentioned by Paul Govaerts in his paper, one should have the feeling that the results provided in 2.3 are the mirror of "overconfidence"! Fortunately (of unfortunately), the daily operations may still give rise to unexpected situations, which may lead to incidents or accidents. Indeed, if it seems that the level of safety has increased, the safety management can really be recognised as an "unceasing process". Let us now emphasize on some supporting events for maintaining a high level of attention and preparedness.

3.1 The radiological risks

Having implemented an ALARA procedure in the early nineties, the main result expected namely, a decrease of the individual as well as the collective dose has been reached in normal operations. Nevertheless, some incidents with potential consequences still have to be indicated.

For instance, bad labelling of transport containers, confusion between radioactive samples and a monitoring device, plumes of smoke generated by a glow lamp foreseen to be used under water but... working in a dry environment are examples of events which still require attention. If their consequences have been negligible, their analysis showed that working with procedures under a quality management system doesn't guarantee the avoidance of deviations of which the consequences could be potentially serious.

3.2 The non radiological risks

These risks are well recognised. With their nature and their occurrence has been dealt with since many years. Data are available and the nuclear sector is also "exposed" to fire, to injuries due to the use of chemical products. But, quite recently, some new observations were made that lead to the need of another approach. We would like to emphasize these trends instead of recalling facts and data already presented in this workshop or more generally, in the literature dealing with the so called "conventional safety".

3.3 Interaction between radiological and non radiological risks

During the decommissioning of the BR3 power plant, we discovered that the asbestos concentration in a controlled area was above the legal limit value. Applying the regulations for asbestos removal and the implementation of radiological protection requirements are not compatible. An in-depth analysis and an ALARA procedure for reducing both the radiological risk and the risk of inhalation of the asbestos required a multidisciplinary approach involving players from different health and safety disciplines.

3.4 Transfer of risks

In order to perform some maintenance work in the controlled area, a worker had to use a ladder. Before entering into the controlled area, he decided to place some plastic sheets around the lower ends of his ladder. These lower ends are made of rubber pieces in order to avoid the ladder to glide. But the worker estimated that these rubber pieces should not be easily cleared from contamination. So, he went to the area, climbed on his ladder and... has to stay at home for three months with a broken leg!

4. THE "UNCEASING" PROCESS

From the examples provided in the former section, keeping in mind the main findings described in section 1, the choice of the word "unceasing" is really justified. Of course, this statement should not appear as surprising for the safety advisors. Nevertheless, there is a problem in this sense that up to the last years, attention was mainly paid to the radiological risks related to radiation sources and practices involving ionising radiation. This approach seems to be no longer acceptable.

Indeed, beside the factors that are already mentioned in another paper (Paul Govaerts), one has to observe an increasingly need for the involvement of all the safety advisors on the "nuclear" workplaces. If the ALARA approach reaches without any doubt, its objectives, there is a need for evolution. For the applications of ionising radiations and nuclear power, the optimisation may not be restricted to the radiological risks.

The role of the Health Physics and Safety Department is to cope with the necessary evolution of the risk management. In its daily operation, this department is faced with changing conditions, changing workers,..., but, under some restrictions (namely, for low level of individual doses), other risks became more significant. The conclusion is that the Health Physics and Safety Department has to involve qualified experts in other fields and that the approach of safety on the workplace must be performed by a team of experts. This new approach, referred to as the ASARA approach by some people, has been considered by the IAEA and by ILO in their International Action Plan as one of the most important issues for the next years.

The conditions for developing such ASARA approach, the involvement of the stakeholders and, last but not least, the perception of risk by the workers should constitute the driving forces of the future works related to the safety at the workplaces.

THE LESSONS OF 48 YEARS' OPERATION OF THE AM RESEARCH REACTOR

(World's First Nuclear Power Plant)

L. A. Kochetkov and V. Ya. Poplavko
RF SSC (Russian Federation State Scientific Center)
Institute of Physics and Power Engineering (RF SSC-IPPE)

Abstract: The world's first nuclear power plant (NPP) was developed and commissioned 50 years ago in the city of Obninsk ; the plant was soon converted to a research reactor. The NPP was developed in a very short period of time – 4 years – while resources were stretched incredibly thin in the effort to restore a shattered wartime economy – industry, the fuel and energy complex, and agriculture. This papers retraces shortly the history of this reactor, and sketches the lessons learned and the experience feedback which have be drawn from it.

F. Lambert and Y. Volkov (eds.), Safety Improvements through Lessons Learned from Operational Experience in Nuclear Research Facilities, 197–202.

1. INTRODUCTION

The world's first nuclear power plant (NPP) was developed and commissioned 50 years ago in the city of Obninsk, at the V Laboratory industrial site (now the State Scientific Center – A. I. Leypunskiy Institute of Physics and Power Engineering); the plant was soon converted to a research reactor.

A Government Resolution for construction of the First NPP was signed in May 1959, and construction began in September 1951 [1].

The world's first NPP was developed in a very short period of time – 4 years – while resources were stretched incredibly thin in the effort to restore a shattered wartime economy – industry, the fuel and energy complex, and agriculture. This was possible because the development of the First NPP was assigned the status of a top-priority political and economic objective.

2. FIRST LESSON LEARNED

Having developed nuclear weapons, hence having provided for adequate defensive capability, national leaders, in building the First NPP, were trying to demonstrate that nuclear arms were not their chosen path. On the other hand, given the postwar devastation of the country, there was an acute hunger for energy, and the successful solution of the problem of nuclear power generation would help to provide energy for industry, transportation and the general public. The national leaders enlisted the top scientists in the country, led by I. V. Kurchatov, head of the "uranium project," to accomplish this task, and the Special Committee under the State Defense Committee headed by L. P. Beria and the First Main Directorate of the national government headed by B. L. Vannikov provided administrative leadership and strict control of the development of the First NPP. The combination of powerful scientific and administrative leadership made possible extremely fast solution of complex scientific, technical and financial problems. This is the first lesson of the successful development of the First NPP.

Another important factor in the successful development of the First NPP was combining the efforts of the venerable scientists of the country, highly qualified engineers and well-trained young specialists. Specialists of the NIIKhIMMASh (Chemical Machine Building Research Institute; now the NIKIET – Power Engineering Research and Design Institute) headed by Professor N. A. Dollezhal and the Leningrad Design Institute headed by A. I. Gutov were recruited to work on the design. I. V. Kurchatov recruited leading scientists from the Institute of Atomic Energy, the Institute of Theoretical and Experimental Physics, the Institute of Physics and Power Engineering and many other institutions to solve specific problems. In March 1951, the scientific leadership of the project to develop the First NPP was transferred to the IPPE under the supervision of its director, D. I. Blokhintsev.

3. SECOND LESSON LEARNED

Training for young specialists was set up in 1946 at special departments of the Moscow Power Engineering Institute and the Moscow Mechanical Institute. In addition to these institutions, Moscow University, the Moscow Physics and Engineering Institute and the Leningrad Polytechnic Institute provided young specialists. Primary attention was devoted to training future operations personnel. In addition to sound theoretical training at the institutes, the future operators acquired practical experience on industrial (military) reactors at Chelyabinsk-40 and the MR reactor built in 1952 at the Institute of Atomic Energy. The training of specialists at all levels for operating the NPP was tested in examinations by a committee made up of leading scientists and engineers. The skills of operations personnel were tested by annual examinations. Subsequently, when several dozen experimental reactors were built in the USSR, conferences were organized under the supervision of Academician A. P. Aleksandrov to exchange experience in the operation of these reactors; conferences were held every 1-2 years.

This kind of concern and supervision over qualified staffs of young specialists on the part of the leading scientists was the foundation for the successful operation of nuclear reactors. Neither the development nor the operation of nuclear reactors would have been possible without prominent scientists in the fields of physics, chemistry, mathematics, materials science, hydraulics and thermodynamics and highly qualified young specialists. This is the second lesson from the experience of the development and operation of the First NPP.

4. SAFETY CONCERN

Nuclear power generation served as the model for new, high-level, potentially hazardous technologies. It required a new approach to the problem of safety at every stage of development of the NPP, from design to operation; a new operating culture acknowledging safety as a priority; and the participation of scientific teams in both the design process and the operation of the NPP. The first offshoots of this safety philosophy can also be traced to the First NPP: the use of feedback; self-protection of the plant; the use of passive protection equipment; the use of a multi-channel approach; and careful training, re-training and monitoring of personnel.

In providing for emergency removal, for example, the design called for two independent electric power systems; and a battery power supply for all safety systems was included to support full shutdown. Redundant reactor power monitoring and control systems were provided. A year before startup of the plant, analysis revealed the potential for reactor runaway in the accidental penetration of water from fuel channel tube failure into the graphite work of the reactor. A positive effect on the reaction rate would develop due to the non-optimal fuel channel grid spacing, and the effect would reach a hazardous level on the assumption of immediate homogeneous filling of all pores in the graphite work of the core. Painstaking analysis and experimental work yielded the following results:

- two systems were developed to monitor water leaks from each fuel channel;
- two systems to remove water and steam from the graphite work were developed and installed, and hydraulic seals were placed on the drainage line and the gas line at the outlet from the reactor vessel;
- it was confirmed experimentally that the hypothesis of immediate homogeneous filling of the graphite work with water is not realized, and operating experience confirmed this conclusion.

Nevertheless, the graphite work was saturated once with moisture absorbed from the air during a prolonged reactor shutdown, but the operators managed to avoid the worst potential consequences on that occasion. In the USSR, and later in Russia as well, in contrast to other countries, a project scientific supervisor was established among the developers in the development and operation of nuclear reactors (power, research and special-purpose reactors). The organization servicing as scientific supervisor for the project, as a rule, was responsible for justifying plant safety, developing analysis codes and normative documents, and solving new problems. Two organizations – the Institute of Atomic Energy and the Institute of Physics and Power Engineering – generally served as scientific supervisors. The Institute of Theoretical and Experimental Physics and the Atomic Reactor Research Institute were scientific supervisors for individual projects. In light of the high potential hazards represented by nuclear reactors, this kind of practice in the management of reactor development work should be considered justified.

5. EXPERIENCE FEEDBACK

Of course, the necessary scope of experimental justification of the project could not be completed in 4 years and had to be dealt with during the operating period. Of the equipment defects identified during operation, two were the most important [2]. The first of these involved the use of gland pumps in the primary circuit, since glandless pumps had not yet been developed at that time. To prevent leaks of hot, radioactive water through the bearings of a pump shaft with a gland seal, cold water was fed into the bearings at a pressure higher than the pressure in the respective pump chamber. This so-called "plug" water was supplied by piston make-up pumps with a power supply, including battery power. If the make-up plug water is not fed into the bearing in some situation, the pump shaft will seize. This happened once during shutdown of the power plant. The operator managed to "get the shaft going" that day and save the unit. By that time, glandless pumps were available and were used to replace the pumps called for in the design.

The second problem involved repeated water leaks through cracks in fuel channel tubes. The water, as it penetrated graphite work heated to 500 - 600°C, caused sudden deterioration in the gas composition: instead of pure helium, CO_2, O_2 and H_2 appeared in significant quantities in the gas composition. This development caused an increase in the graphite temperature to 700 - 730°C, accelerated graphite "burnout," and, most dangerous of all, the risk of detonation of the explosive mixture. Studies performed on defective pipe segments in a hot chamber established that cracks develop as a result of corrosion under stress. To prevent a recurrence of such events, the fuel channel design and the system for feeding gas into the core were modified.

After the flaws discovered during the initial operating period had been corrected, the plant operating mode was stabilized, the rated power level was achieved, and the unit was converted to a research reactor, although thermal energy recycling and isotope production continued.

A total of 18 loops were created for research on the AM reactor. Numerous programs were carried out on these loops, of which the main programs were:

- complex studies of startup and transient modes of the first two power-generating units of the Beloyarsk NPP, with superheating of steam inside the reactor to 510°C;
- complex studies of startup and transient modes of power-generating units of the Bilibino Nuclear Central Heating and Power Plant with natural coolant circulation;
- tests of fuel elements of the Beloyarsk and Bilibino NPPs, the TES-3 NPP and the power-generating channels of BUK and TOPAZ reactor plants for space purposes.

The operation of the reactor unit was halted in 2002, after 48 years of operation.

Work is currently being done on the unit for a decommissioning program. The experience gained in this new work will be yet another contribution of the world's first NPP to the common body of knowledge on research reactors.

6. REFERENCES

1. Kudinova, L. I., and A. V. Shchegelskiy. Towards a History of the Peaceful Use of Nuclear Power in the USSR. 1944-1951 (Documents and Materials). Obninsk, GNTs RF-FEI (RF SSC-IPPE), 1994, pp. XV – XVIII, 8.
2. Kochetkov, L. A. The World's First NPP and Its Role in the Development of Nuclear Power.

SUMMARY OF THE GENERAL DISCUSSIONS

Notes collected by

Christine FELTIN
Commissariat à l'Energie Atomique, INSTN

1. 1ST GENERAL DISCUSSION

This discussion focused on: *" Lessons Learned and Defence in Depth in Nuclear Research Facilities"*.

A summary of the exchanges is reported below :

- Since many events are connected to the Human Factor, it is important to :
  - Organize adequate training, and assess the qualifications,
  - Find money for training, and training equipment like simulators,
  - Draw the lessons learned from errors and mistakes to minimize their number.

- Because our systems are not ideal, and the human reliability is limited, it is necessary to have sturdy systems,

- To detect any minor miss, it is necessary to develop the safety culture of our operating teams, but also to foster the "*reporting without complex*" attitude. How can we do that?

- In most real incident it is impossible to say: "this is human failure, this is equipment failure" because factors are more or less entangled. Safety culture and human factor have special place in experience feedback :

F. Lambert and Y. Volkov (eds.), Safety Improvements through Lessons Learned from Operational Experience in Nuclear Research Facilities, 203–208.

  - For engineers, the analysis of technical failures is easy, that of human failure is difficult. The solution in CEA is to offer the management the assistance of a consultant, who guarantees secrecy ;
  - Staff safety culture is indeed important, but also that of management, as they have to decide which resources will be spent to cope with a problem that caused an incident.

- All three are important : Man, Technology, Organization. The top management safety culture is important as they are responsible for the organization and good communication between all those involved in the facility's operation.

- For particular types of operation, like decommissioning where private contractors are involved, their perception about reporting on incidents is negative, because it could adversely affect their having contracts. We attempt to overcome this by applying a tolerant culture and take the positive aspects, and encourage them to report short faults in safety performance reviews and reports, and to foster free reporting.

- A safety committee could be set up to analyse errors and incidents. It is necessary to set up the basis for good analysis: experts from outside should work hand in hand with operators at the reactor.

- The harmonization of the reporting, selection, and rating of incidents could be beneficial. The INES (International Nuclear Events Scale) is not yet an official system and even if all INES level 2 incidents are reported, the situation is different among countries for levels 1 and 0. This workshop could propose some actions about harmonization between systems. The weak point in incident analysis is reporting as generally only one communication line exists: a system should be set up to establish a second communication line.

- Quality Assurance is needed at each step: detection, analysis, decision of what to correct and following of the actions plan taken. The first steps are easy, but quality system involving the decisions is quite difficult and expensive. The last step is the most important. Until now we have been working more on the analysis ; we should work more on the corrective actions and their implementation.

- In this workshop, an interesting idea is prevention of major incidents based on the analysis of minor ones. It is necessary to define minor incidents (*the incidents rated zero or one on the INES scale*). Minor incidents happen more frequently; the analysis of these minor incidents is like monitoring reliability.

2. FINAL GENERAL DISCUSSION

The principal conclusions reached, the main ideas, comments, and practices to promote which were stressed by the key speakers and the participants during the final general discussion, are summarized below :

- Minor incidents should not be considered as unimportant because they have no real consequences, but as a very interesting source of information to maintain or improve the safety level of our Nuclear Research Facilities. The analysis of minor events is important to allow for preventive actions to be taken before curative actions become necessary.

- The role of the higher management of the facility is essential to promote, in the facilities operating teams, the feeling that it is most valuable :

  - to detect these minor events, or near misses,
  - to report as soon as possible to the management,
  - to perform further analysis in terms of potential consequences, and of test of the strength or weaknesses of the lines of defence against more severe hazards.

- A great proportion of small events are related to Human Factor :
  - Poor qualification,
  - Need for Education and training, including continuous training,
  - Nuclear not attractive due to very low salaries in some countries,

  - Physics, engineering not very popular among students (slowly improving but not as fast as we would like).

- Western and Eastern colleagues have similar problems: we feel closer to each other, and we need to meet and exchange more frequently. Attention to research units and to safety is a very important matter. Attention was also focused upon these by IAEA since a great number of research reactors have to be decommissioned, and there is a generic safety problem. The economic situation is difficult.

- Important lessons learned in this ARW :
  - to focus upon near misses;
  - research, both theoretical and experimental could be of particular importance on how to improve analysis of minor events, with the perspective of creation of models, for prediction of the probability of the small accident transformation into bigger accident if no proper measures would be taken, and for issuing the recommendations for the necessary measures to prevent such transformation

- Some participants wish further information, and exchange, on methodology and procedures for minor events analysis, which could apply to a wide range of various facilities.

- Notable similarities between west and Russia:
  - Good practices,
  - Communication with trade unions : safety representatives speak to the work force,
  - Regular committees,
- UK is looking at Human Factor in procedure deviations: how to informally improve procedures.

- Multiple aspects were presented during the ARW :
  - human error,
  - importance of the lessons learned from incidents to prevent more severe accidents,
  - need to build a confident atmosphere in the organisation: the management must be open to discussion, to rapidly obtain information on minor incidents ;
  - Enough money to proper maintenance is necessary ;
  - Good relationship with authority ;

- Suggestions to improve our treatment of minor incidents have been proposed.

- People will make less error if they understand the reasons behind the procedures:
  - they must have a knowledge beyond what is needed to apply the procedures and do the job;
  - The information upon minor incidents must be shared : it is always easier to discuss incidents that occurred in another place ;
  - Social aspects must be recognized even if they are difficult to address.

- To use the lessons of experience we must have:
  - cultural elements :
    - skills, good training,
    - safety culture, that is when staff is ready to report to the boss and the boss is happy with the report,
  - practical elements : learning and implementing the lessons from experience requires time and money, and a good organization to set appropriate priorities;
  - technical elements : because of human factors, we must keep the general idea of sturdy facilities. We don't want to put on the operator the consequences of design weakness or of lack of money for maintenance.

- It is important :
  - not only to observe what failed but also what operated properly : the effectiveness of the different defence lines ;
  - take into account other tools: multi-disciplinary exchanges, peer exchanges, Human Factor expertise ;
  - take into account any type of events or risks and not only nuclear safety events and perform global assessment,: we have to live very strong social and structural changes, we have to imagine how to cope with safety in this context.

- ALARA (as low as reasonably achievable) and ASARA (as safe as reasonably achievable) networks exist: but only within European Union countries. How to motivate other countries to participate and establish collecting processes for all this information.

- Regulatory practices: there are some differences between our practices, but the reactor regulatory framework has many similarities between France and Russia. We want to take care of minor misses in the licensing process. Regulation must be well enforced. A good commitment of operators and authority is necessary. French Nuclear Safety Authority is now experiencing to include radiation protection.

GPSR Compliance

The European Union's (EU) General Product Safety Regulation (GPSR) is a set of rules that requires consumer products to be safe and our obligations to ensure this.

If you have any concerns about our products, you can contact us on ProductSafety@springernature.com

In case Publisher is established outside the EU, the EU authorized representative is:

Springer Nature Customer Service Center GmbH
Europaplatz 3
69115 Heidelberg, Germany

Batch number: 10761599

Printed by Printforce, the Netherlands